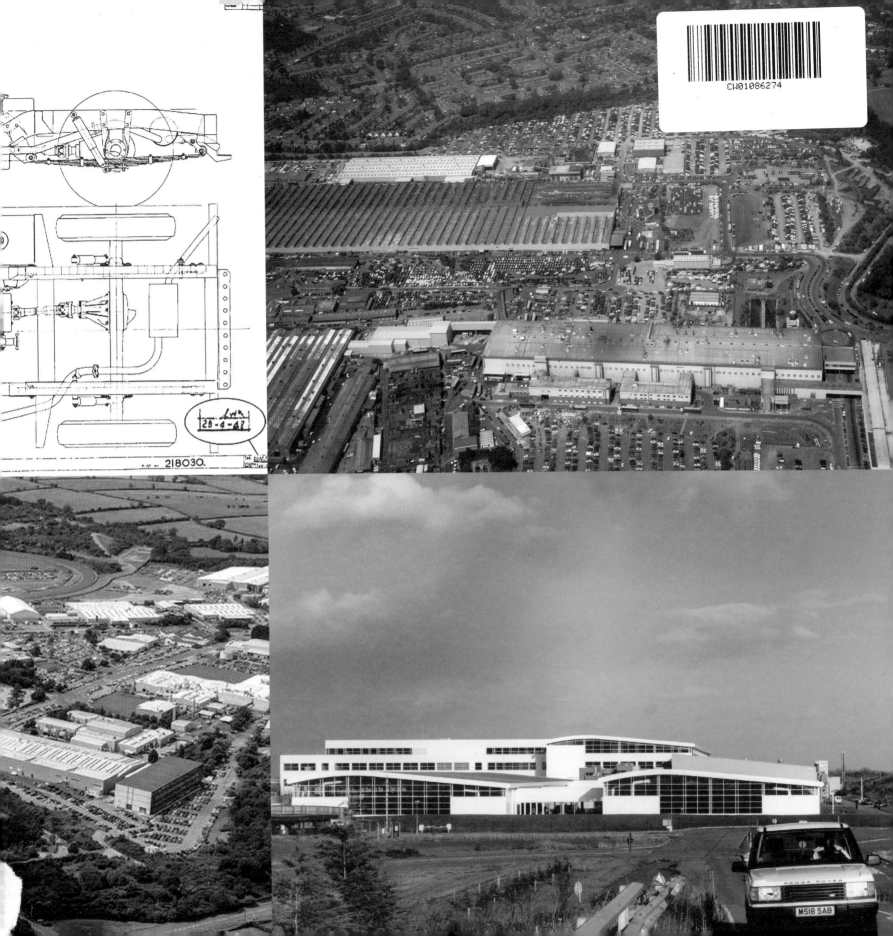

Land Rover
DESIGN
— 70 years of success

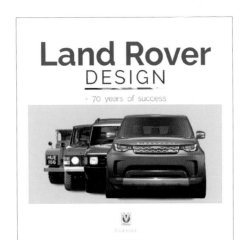

Also from Veloce Publishing –

11⁄2-litre GP Racing 1961-1965 (Whitelock)
AC Two-litre Saloons & Buckland Sportscars (Archibald)
Alfa Romeo 155/156/147 Competition Touring Cars (Collins)
Alfa Romeo Giulia Coupé GT & GTA (Tipler)
Alfa Romeo Montreal – The dream car that came true (Taylor)
Alfa Romeo Montreal – The Essential Companion (Classic Reprint of 500 copies) (Taylor)
Alfa Tipo 33 (McDonough & Collins)
Alpine & Renault – The Development of the Revolutionary Turbo F1 Car 1968 to 1979 (Smith)
Alpine & Renault – The Sports Prototypes 1963 to 1969 (Smith)
Alpine & Renault – The Sports Prototypes 1973 to 1978 (Smith)
Anatomy of the Classic Mini (Huthert & Ely)
Anatomy of the Works Minis (Moylan)
Armstrong-Siddeley (Smith)
Art Deco and British Car Design (Down)
Autodrome (Collins & Ireland)
Autodrome 2 (Collins & Ireland)
Automotive A-Z, Lane's Dictionary of Automotive Terms (Lane)
Automotive Mascots (Kay & Springate)
Bahamas Speed Weeks, The (O'Neil)
Bentley Continental, Corniche and Azure (Bennett)
Bentley MkVI, Rolls-Royce Silver Wraith, Dawn & Cloud/Bentley R & S-Series (Nutland)
Bluebird CN7 (Stevens)
BMC Competitions Department Secrets (Turner, Chambers & Browning)
BMW 5-Series (Cranswick)
BMW Z-Cars (Taylor)
BMW Boxer Twins 1970-1995 Bible, The (Falloon)
BMW Cafe Racers (Cloesen)
BMW Custom Motorcycles – Choppers, Cruisers, Bobbers, Trikes & Quads (Cloesen)
BMW – The Power of M (Vivian)
Bonjour – Is this Italy? (Turner)
British 250cc Racing Motorcycles (Pereira)
British at Indianapolis, The (Wagstaff)
British Café Racers (Cloesen)
British Cars, The Complete Catalogue of, 1895-1975 (Culshaw & Horrobin)
British Custom Motorcycles – The Brit Chop – choppers, cruisers, bobbers & trikes (Cloesen)
BRM – A Mechanic's Tale (Salmon)
BRM V16 (Ludvigsen)
BSA Bantam Bible, The (Henshaw)
BSA Motorcycles – the final evolution (Jones)
Bugatti Type 40 (Price)
Bugatti 46/50 Updated Edition (Price & Arbey)
Bugatti T44 & T49 (Price & Arbey)
Bugatti 57 2nd Edition (Price)
Bugatti Type 57 Grand Prix – A Celebration (Tomlinson)
Caravan, Improve & Modify Your (Porter)
Caravans, The Illustrated History 1919-1959 (Jenkinson)
Caravans, The Illustrated History From 1960 (Jenkinson)
Carrera Panamericana, La (Tipler)
Car-tastrophes – 80 automotive atrocities from the past 20 years (Honest John, Fowler)
Chrysler 300 – America's Most Powerful Car 2nd Edition (Ackerson)
Chrysler PT Cruiser (Ackerson)
Citroën DS (Bobbitt)
Classic British Car Electrical Systems (Astley)
Cobra – The Real Thing! (Legate)
Competition Car Aerodynamics 3rd Edition (McBeath)
Competition Car Composites A Practical Handbook (Revised 2nd Edition) (McBeath)
Concept Cars, How to illustrate and design (Dewey)
Cortina – Ford's Bestseller (Robson)
Cosworth – The Search for Power (6th edition) (Robson)
Coventry Climax Racing Engines (Hammill)
Daily Mirror 1970 World Cup Rally 40, The (Robson)
Daimler SP250 New Edition (Long)
Datsun Fairlady Roadster to 280ZX – The Z-Car Story (Long)
Dino – The V6 Ferrari (Long)
Dodge Challenger & Plymouth Barracuda (Grist)
Dodge Charger – Enduring Thunder (Ackerson)
Dodge Dynamite! (Grist)
Dorset from the Sea – The Jurassic Coast from Lyme Regis to Old Harry Rocks photographed from its best viewpoint (also Souvenir Edition) (Belasco)
Draw & Paint Cars – How to (Gardiner)
Drive on the Wild Side, A – 20 Extreme Driving Adventures From Around the World (Weaver)
Ducati 750 Bible, The (Falloon)
Ducati 750 SS 'round-case' 1974, The Book of the (Falloon)
Ducati 860, 900 and Mille Bible, The (Falloon)
Ducati Monster Bible (New Updated & Revised Edition), The (Falloon)
Ducati 916 (updated edition) (Falloon)
Dune Buggy, Building A – The Essential Manual (Shakespeare)
Dune Buggy Files (Hale)
Dune Buggy Handbook (Hale)
East German Motor Vehicles in Pictures (Suhr/Weinreich)
Fast Ladies – Female Racing Drivers 1888 to 1970 (Bouzanquet)
Fate of the Sleeping Beauties, The (op de Weegh/Hottendorff/op de Weegh)
Ferrari 288 GTO, The Book of the (Sackey)
Ferrari 333 SP (O'Neil)
Fiat & Abarth 124 Spider & Coupé (Tipler)
Fiat & Abarth 500 & 600 – 2nd Edition (Bobbitt)
Fiats, Great Small (Ward)
Fine Art of the Motorcycle Engine, The (Peirce)
Ford Cleveland 335-Series V8 engine 1970 to 1982 – The Essential Source Book (Hammill)
Ford F100/F150 Pick-up 1948-1996 (Ackerson)
Ford F150 Pick-up 1997-2005 (Ackerson)
Ford GT – Then, and Now (Streather)
Ford GT40 (Legate)
Ford Midsize Muscle – Fairlane, Torino & Ranchero (Cranswick)
Ford Model Y (Roberts)
Ford Small Block V8 Racing Engines 1962-1970 – The Essential Source Book (Hammill)
Ford Thunderbird From 1954, The Book of the (Long)
Formula 5000 Motor Racing, Back then ... and back now (Lawson)
Forza Minardi! (Vigar)
France: the essential guide for car enthusiasts – 200 things for the car enthusiast to see and do (Parish)
From Crystal Palace to Red Square – A Hapless Biker's Road to Russia (Turner)
Funky Mopeds (Skelton)
Grand Prix Ferrari – The Years of Enzo Ferrari's Power, 1948-1980 (Pritchard)
Grand Prix Ford – DFV-powered Formula 1 Cars (Robson)
GT – The World's Best GT Cars 1953-73 (Dawson)
Hillclimbing & Sprinting – The Essential Manual (Short & Wilkinson)
Honda NSX (Long)
Inside the Rolls-Royce & Bentley Styling Department – 1971 to 2001 (Hull)
Intermeccanica – The Story of the Prancing Bull (McCredie & Reisner)
Italian Cafe Racers (Cloesen)
Italian Custom Motorcycles (Cloesen)
Jaguar, The Rise of (Price)
Jaguar XJ 220 – The Inside Story (Moreton)
Jaguar XJ-S, The Book of the (Long)
Japanese Custom Motorcycles – The Nippon Chop – Chopper, Cruiser, Bobber, Trikes and Quads (Cloesen)
Jeep CJ (Ackerson)
Jeep Wrangler (Ackerson)
The Jowett Jupiter – The car that leaped to fame (Nankivell)
Karmann-Ghia Coupé & Convertible (Bobbitt)
Kawasaki Triples Bible, The (Walker)
Kawasaki Z1 Story, The (Sheehan)
Kris Meeke – Intercontinental Rally Challenge Champion (McBride)
Lamborghini Miura Bible, The (Sackey)
Lamborghini Urraco, The Book of the (Landsem)
Lambretta Bible, The (Davies)
Lancia 037 (Collins)
Lancia Delta HF Integrale (Blaettel & Wagner)
Land Rover Series III Reborn (Porter)
Land Rover, The Half-ton Military (Cook)
Laverda Twins & Triples Bible 1968-1986 (Falloon)
Lea-Francis Story, The (Price)
Le Mans Panoramic (Ireland)
Lexus Story, The (Long)
Little book of microcars, the (Quellin)
Little book of smart, the – New Edition (Jackson)
Little book of trikes, the (Quellin)
Lola – The Illustrated History (1957-1977) (Starkey)
Lola – All the Sports Racing & Single-seater Racing Cars 1978-1997 (Starkey)
Lola T70 – The Racing History & Individual Chassis Record – 4th Edition (Starkey)
Lotus 18 Colin Chapman's U-turn (Whitelock)
Lotus 49 (Oliver)
Marketingmobiles, The Wonderful Wacky World of (Hale)
Maserati 250F In Focus (Pritchard)
Mazda MX-5/Miata 1.6 Enthusiast's Workshop Manual (Grainger & Shoemark)
Mazda MX-5/Miata 1.8 Enthusiast's Workshop Manual (Grainger & Shoemark)
Mazda MX-5 Miata, The book of the – The 'Mk1' NA-series 1988 to 1997 (Long)
Mazda MX-5 Miata Roadster (Long)
Mazda Rotary-engined Cars (Cranswick)
Maximum Mini (Booij)
Meet the English (Bowie)
Mercedes-Benz SL – R230 series 2001 to 2011 (Long)
Mercedes-Benz SL – W113-series 1963-1971 (Long)
Mercedes-Benz SL & SLC – 107-series 1971-1989 (Long)
Mercedes-Benz SLK – R170 series 1996-2004 (Long)
Mercedes-Benz SLK – R171 series 2004-2011 (Long)
Mercedes-Benz W123-series – All models 1976 to 1986 (Long)
Mercedes G-Wagen (Long)
MGA (Price Williams)
MGB & MGB GT– Expert Guide (Auto-doc Series) (Williams)
MGB Electrical Systems Updated & Revised Edition (Astley)
Micro Caravans (Jenkinson)
Micro Trucks (Mort)
Microcars at Large! (Quellin)
Mini Cooper – The Real Thing! (Tipler)
Mini Minor to Asia Minor (West)
Mitsubishi Lancer Evo, The Road Car & WRC Story (Long)
Montlhéry, The Story of the Paris Autodrome (Boddy)
Morgan Maverick (Lawrence)
Morgan 3 Wheeler – back to the future!, The (Dron)
Morris Minor, 60 Years on the Road (Newell)
Moto Guzzi Sport & Le Mans Bible, The (Falloon)
Motor Movies – The Posters! (Veysey)
Motor Racing – Reflections of a Lost Era (Carter)
Motor Racing – The Pursuit of Victory 1930-1962 (Carter)
Motor Racing – The Pursuit of Victory 1963-1972 (Wyatt/Sears)
Motor Racing Heroes – The Stories of 100 Greats (Newman)
Motorcycle Apprentice (Cakebread)
Motorcycle GP Racing in the 1960s (Pereira)
Motorcycle Road & Racing Chassis Designs (Noakes)
Motorhomes, The Illustrated History (Jenkinson)
Motorsport In colour, 1950s (Wainwright)
MV Agusta Fours, The book of the classic (Falloon)
N.A.R.T. – A concise history of the North American Racing Team 1957 to 1983 (O'Neil)
Nissan 300ZX & 350Z – The Z-Car Story (Long)
Nissan GT-R Supercar: Born to race (Gorodji)
Northeast American Sports Car Races 1950-1959 (O'Neil)
The Norton Commando Bible – All models 1968 to 1978 (Henshaw)
Nothing Runs – Misadventures in the Classic, Collectable & Exotic Car Biz (Slutsky)
Off-Road Giants! (Volume 1) – Heroes of 1960s Motorcycle Sport (Westlake)
Off-Road Giants! (Volume 2) – Heroes of 1960s Motorcycle Sport (Westlake)
Off-Road Giants! (volume 3) – Heroes of 1960s Motorcycle Sport (Westlake)
Pass the Theory and Practical Driving Tests (Gibson & Hoole)
Peking to Paris 2007 (Young)
Pontiac Firebird – New 3rd Edition (Cranswick)
Porsche Boxster (Long)
Porsche 356 (2nd Edition) (Long)
Porsche 908 (Födisch, Neßhöver, Roßbach, Schwarz & Roßbach)
Porsche 911 Carrera – The Last of the Evolution (Corlett)
Porsche 911R, RS & RSR, 4th Edition (Starkey)
Porsche 911, The Book of the (Long)
Porsche 911 – The Definitive History 2004-2012 (Long)
Porsche – The Racing 914s (Smith)
Porsche 911SC 'Super Carrera' – The Essential Companion (Streather)
Porsche 914 & 914-6: The Definitive History of the Road & Competition Cars (Long)
Porsche 924 (Long)
The Porsche 924 Carreras – evolution to excellence (Smith)
Porsche 928 (Long)
Porsche 944 (Long)
Porsche 964, 993 & 996 Data Plate Code Breaker (Streather)
Porsche 993 'King Of Porsche' – The Essential Companion (Streather)
Porsche 996 'Supreme Porsche' – The Essential Companion (Streather)
Porsche 997 2004-2012 – Porsche Excellence (Streather)
Porsche Racing Cars – 1953 to 1975 (Long)
Porsche Racing Cars – 1976 to 2005 (Long)
Porsche – The Rally Story (Meredith)
Porsche: Three Generations of Genius (Meredith)
Preston Tucker & Others (Linde)
RAC Rally Action! (Gardiner)
RACING COLOURS – MOTOR RACING COMPOSITIONS 1908-2009 (Newman)
Racing Line – British motorcycle racing in the golden age of the big single (Guntrip)
Rallye Sport Fords: The Inside Story (Moreton)
Renewable Energy Home Handbook, The (Porter)
Roads with a View – England's greatest views and how to find them by road (Corfield)
Rolls-Royce Silver Shadow/Bentley T Series Corniche & Camargue – Revised & Enlarged Edition (Bobbitt)
Rolls-Royce Silver Spirit, Silver Spur & Bentley Mulsanne 2nd Edition (Bobbitt)
Rootes Cars of the 50s, 60s & 70s – Hillman, Humber, Singer, Sunbeam & Talbot (Rowe)
Rover P4 (Bobbitt)
Runways & Racers (O'Neil)
Russian Motor Vehicles – Soviet Limousines 1930-2003 (Kelly)
Russian Motor Vehicles – The Czarist Period 1784 to 1917 (Kelly)
RX-7 – Mazda's Rotary Engine Sports car (Updated & Revised New Edition) (Long)
Scooters & Microcars, The A-Z of Popular (Dan)
Scooter Lifestyle (Grainger)
SCOOTER MANIA! – Recollections of the Isle of Man International Scooter Rally (Jackson)
Singer Story: Cars, Commercial Vehicles, Bicycles & Motorcycle (Atkinson)
Sleeping Beauties USA – abandoned classic cars & trucks (Marek)
SM – Citroën's Maserati-engined Supercar (Long & Claverol)
Speedway – Auto racing's ghost tracks (Collins & Ireland)
Sprite Caravans, The Story of (Jenkinson)
Standard Motor Company, The Book of the (Robson)
Steve Hole's Kit Car Cornucopia – Cars, Companies, Stories, Facts & Figures: the UK's kit car scene since 1949 (Hole)
Subaru Impreza: The Road Car And WRC Story (Long)
Supercar, How to Build your own (Thompson)
Tales from the Toolbox (Oliver)
Tatra – The Legacy of Hans Ledwinka, Updated & Enlarged Collector's Edition of 1500 copies (Margolius & Henry)
Taxi! The Story of the 'London' Taxicab (Bobbitt)
To Boldly Go – twenty six vehicle designs that dared to be different (Hull)
Toleman Story, The (Hilton)
Toyota Celica & Supra, The Book of Toyota's Sports Coupés (Long)
Toyota MR2 Coupés & Spyders (Long)
Triumph Bonneville Bible (59-83) (Henshaw)
Triumph Bonneville!, Save the – The inside story of the Meriden Workers' Co-op (Rosamond)
Triumph Motorcycles & the Meriden Factory (Hancox)
Triumph Speed Twin & Thunderbird Bible (Woolridge)
Triumph Tiger Cub Bible (Estall)
Triumph Trophy Bible (Woolridge)
Triumph TR6 (Kimberley)
TT Talking – The TT's most exciting era – As seen by Manx Radio TT's lead commentator 2004-2012 (Lambert)
Two Summers – The Mercedes-Benz W196R Racing Car (Ackerson)
TWR Story, The – Group A (Hughes & Scott)
Unraced (Collins)
Velocette Motorcycles – MSS to Thruxton – New Third Edition (Burris)
Vespa – The Story of a Cult Classic in Pictures (Uhlig)
Vincent Motorcycles: The Untold Story since 1946 (Guyony & Parker)
Volkswagen Bus Book, The (Bobbitt)
Volkswagen Bus or Van to Camper, How to Convert (Porter)
Volkswagens of the World (Glen)
VW Beetle Cabriolet – The full story of the convertible Beetle (Bobbitt)
VW Beetle – The Car of the 20th Century (Copping)
VW Bus – 40 Years of Splitties, Bays & Wedges (Copping)
VW Bus Book, The (Bobbitt)
VW Golf: Five Generations of Fun (Copping & Cservenka)
VW – The Air-cooled Era (Copping)
VW T5 Camper Conversion Manual (Porter)
VW Campers (Copping)
Volkswagen Type 3, The book of the – Concept, Design, International Production Models & Development (Glen)
You & Your Jaguar XK8/XKR – Buying, Enjoying, Maintaining, Modifying – New Edition (Thorley)
Which Oil? – Choosing the right oils & greases for your antique, vintage, veteran, classic or collector car (Michell)
Works Minis, The Last (Purves & Brenchley)
Works Rally Mechanic (Moylan)

www.veloce.co.uk

Also available ... Apps Download now from www.digital.veloce.co.uk

Also available ... eBook Download now from www.digital.veloce.co.uk

Belvedere Hubble & Hattie Kids Hubble & Hattie

First published in August 2018 by Veloce Publishing Limited, Veloce House, Parkway Farm Business Park, Middle Farm Way, Poundbury, Dorchester DT1 3AR, England.
Fax 01305 250479 / Tel 01305 260068 / e-mail info@veloce.co.uk / web www.veloce.co.uk or www.velocebooks.com.
ISBN: 978-1-845849-87-0 UPC: 6-36847-04987-4
© 2018 Nick Hull and Veloce Publishing. All rights reserved. With the exception of quoting brief passages for the purpose of review, no part of this publication may be recorded, reproduced or transmitted by any means, including photocopying, without the written permission of Veloce Publishing Ltd. Throughout this book logos, model names and designations, etc, have been used for the purposes of identification, illustration and decoration. Such names are the property of the trademark holder as this is not an official publication. Readers with ideas for automotive books, or books on other transport or related hobby subjects, are invited to write to the editorial director of Veloce Publishing at the above address.
British Library Cataloguing in Publication Data – A catalogue record for this book is available from the British Library. Typesetting, design and page make-up all by Veloce Publishing Ltd on Apple Mac.
Printed in India by Parksons Graphics.

Land Rover
DESIGN
– 70 years of success

NICK HULL

Contents

Foreword by Gerry McGovern 5

Introduction and acknowledgements 6

Chapter 1: Early days of Land Rover design 8

Chapter 2: David Bache era and Range Rover 30

Chapter 3: Independence and expansion of the range 66

Chapter 4: Canley studio and a new owner – BMW 95

Chapter 5: Geoff Upex era and Gaydon studio 120

Chapter 6: The Ford years 142

Chapter 7: Gerry McGovern takes charge 169

Chapter 8: Rapid expansion of design activities 186

Chapter 9: Current Land Rover design 212

Appendix 1: Land Rover model code numbers 231

Appendix 2: Glossary of design terms 233

Index 239

Foreword

Gerry McGovern, Director of Design and Chief Creative Officer, Land Rover

Land Rover is an iconic British brand that is on a journey of transformation. As its custodians, we are creating a new range of highly desirable and incredibly capable SUVs for a world that has changed almost beyond recognition since Land Rover began in 1948. I believe it will do so again in the next 25 years.

This is a fascinating time to be a designer, because so many new ideas and innovations are coming into play, such as autonomous driving, electrification, connectivity, shared ownership and mass urbanisation. The designs we create will be influenced by these factors – which is where the opportunity lies. The most successful companies will retain the essence of their DNA, but shape it in a way that remains relevant to customers. For me, new design solutions must preserve the character of the brand.

Automotive design is a balance between product and industrial disciplines, which of course must be commercially viable. The aesthetic is a fundamental driver for consumers: it's about an emotional connection. Design makes this connection, particularly once technology, reliability and quality become comparable.

Designing a vehicle is a hugely complex and intellectual challenge with so many variable requirements to be considered. Land Rover's unique proposition is the combination of design leadership and engineering integrity. Range Rover customers are very clear about how they want their vehicle to evolve: 'Don't change it, just make it better.' For brand new vehicles such as the Range Rover Evoque and more recently the Velar, design sets the vision and defines the overall package, and then we work with our engineering colleagues to realise that vision.

The new Defender will meet those same exacting standards. It is the most anticipated Land Rover, but this does not change our approach or put pressure on the designers and engineers working on it. While everyone is utterly dedicated to Defender, we approach all our exciting potential vehicles with the same passion.

Land Rover's tireless commitment to offering experiences that customers love for life is at the core of everything we do, and as we celebrate our 70th year we continue to create vehicles that are universally desirable while maintaining the essence of our unique brand.

Introduction and acknowledgements

Despite the prevalence of books on Land Rover, there has never been a book dedicated to the subject of Land Rover design and styling. With the 70th anniversary of the marque in 2018, it seems timely to produce such a book.

This book focusses on the people, projects and process of design, and how this has changed over time. There is a fourth 'p' as well – places. Land Rover has had five design studios in the 70 years since its beginnings in Solihull, and the history of the earlier studios is barely recorded. The identification and description of each of those studios and locations where the photos of various models were taken is one aspect of this book that has never been attempted before, and forms an important documentation of the company history, particularly since many of those locations and buildings no longer exist.

As a profession, automotive design has changed hugely over the past 70 years, and continues to evolve as new processes and methods are developed. In the early days there was little formal training available for new recruits, and designers came from a variety of backgrounds. Up to the 1980s, David Bache and his successors struggled to recruit good designers with the right creative and drawing skills, which led to their involvement with the Royal College of Art in London, and also with Coventry University. Nowadays, the UK is a strong source of design talent in Europe.

Land Rover under David Bache was a forward-looking studio that eschewed whimsical styling trends in favour of good industrial design practice. Many of the early designers working at Rover came from an industrial or jewellery design background, and this was reflected in their attention to ergonomics and striving for beautifully detailed design, such as the original Range Rover.

It was also a forward-thinking company that recruited design graduates when the rest of the UK car industry rather shunned them. Nowadays, young designers require a degree in Automotive Design (plus a stunning portfolio) as a basic requirement for entry into the profession, and women form a key part of any design team, too – and not just in the typical area of colour and trim. Design studios today offer a full career structure, not just for the designers but for clay and digital modellers, also technicians, with migration of staff across companies and continents being commonplace.

Whilst researching this book, it became obvious that the preservation of design material for the archives is of huge value. Each time the design department moved, lots of material was

Introduction and acknowledgements

cleared out and potentially interesting sketches were inevitably lost. Not everything should be saved, of course, but it is important to document the key stages in any project in order for future generations of historians to make sense of it. Likewise, I have sought out rare photos of sketches, models and early prototypes wherever possible, rather than simply using routine images of production models.

Finally, a note on the use of terms. The book adopts the terms used within design and engineering, which are a mix of American and English terminology, chiefly due to the influence of the Pressed Steel company in the 1940s. Thus, 'fender' is the term used rather than 'wing,' and 'trunk' rather than 'boot' because these are the descriptions used on all engineering drawings and in design discussions, not just at Land Rover, but within most car companies globally.

Acknowledgements

I would like to thank Gerry McGovern, Richard Woolley, Dave Saddington, Massimo and Amy Frascella, Phil Simmons, Andy Wheel, Alan Sheppard, David and Kim Brisbourne, Mark Butler, Oliver le Grice, Don Wyatt, Mick Jones, Alan Mobberley, Peter Crowley, Geoff Purkis, Chris Wade, Norman Morris, David Browne and John Stark for finding the time to be interviewed about their recollections and providing detailed information that allowed me to piece together the full story of the design studios.

The vast majority of images were supplied by the JLR PR team, with special thanks to Angela Powell and Lydia Heynes for their efforts in sorting through the older photo archives to come up with a good selection of images from the design files. Other historical photos were supplied by BMIHT archives. Special thanks also to Geoff Upex, Mike Sampson, Maureen Hill, and Land Rover guru Roger Crathorne for providing valuable help with images and information, and to Paul Owen for facilitating access to Gaydon staff.

I am also indebted to David Evans and Land Rover author James Taylor, who both provided a lot of detailed background information while writing the book and supplied a number of additional images from their own collections.

Finally, thanks to the staff at Veloce who have supported the publishing of the book and created a great layout for it.

Nick Hull

1947-1959

Chapter 1

Early days of Land Rover design

In 2015 Land Rover created the largest sand drawing ever produced in the UK. A Defender outline measuring 1km across was drawn on the beach at Red Wharf Bay in Anglesey using a fleet of six Land Rovers. The unique image was a tribute to the moment in 1947 when the Engineering Director of Rover, Maurice Wilks, first sketched the shape for the original Land Rover in the sand of Red Wharf Bay and proposed the idea to his brother Spencer, Rover's Managing Director.

"My father met his brother on the beach at Red Wharf Bay and made a drawing in the sand of how he thought the Land Rover could be made," said Stephen Wilks, son of Maurice. "That was the start of it all, the conception of Land Rover."

Red Wharf Bay sand image, drawn in 2015.

Conception of the Land Rover

The Rover Company had rather a chequered career in its early days, and by the late 1920s the company, based in Coventry, was nearly bankrupt. Brothers Spencer and Maurice Wilks came on board from 1929 and revived the company's fortunes through careful management and the creation of a series of high quality cars that appealed to the professional middle-classes. Maurice was Chief Engineer, and responsible for designing a series of saloons throughout the 1930s that combined conservative looks with excellent build quality and good performance from their overhead-valve engines, while Spencer was Managing Director.

The Rover Company was based in Helen Street, Coventry, some two miles north of the city centre near Foleshill. This was the main production plant, but it also had a small plant in Tyseley, Birmingham for engine build. In 1936, Rover was invited by the government to

Lode Lane plant, 1940s. Rover wisely bought up 200 acres of surrounding agricultural land that was to prove very useful later on as the plant expanded. The wood behind is called Billesley Wood – still in existence today as the site of Land Rover's off-road experience. Note the H-shaped buildings for aero engine testing on the left of the picture. (Courtesy Roger Crathorne)

manage one of the new 'shadow factories' that were being built in anticipation of a potential war with Germany. This was in Acocks Green, Birmingham, and the factory was up and running by July 1937, producing aeroplane engine components.

In April 1939, the Rover board reported that discussions had taken place with the Air Ministry to construct a second shadow factory for aero engines at Lode Lane, south-west of Elmdon aerodrome – now Birmingham airport. This would become the company's main assembly plant and headquarters after the war. Building work for this much larger factory on the northern outskirts of Solihull began in 1939, and was completed for September 1940. During the war it employed 7000 people making Hercules 1700bhp 14-cylinder air-cooled radial aero engines used in Short Stirling heavy bombers that were assembled at Austin's Longbridge plant across the city.

Land Rover – 70 years of success

The engines were made in conjunction with Rootes Group, who had another shadow factory at Ryton on Dunsmore. The first engines were produced in October 1940.

One month later a German raid on Coventry destroyed most of the city centre, including the Helen Street works, whereupon the Rover design team relocated to Chesford Grange hotel, out towards Leamington Spa. By 1944 Rover had 24,000 employees in 18 factories around the UK, including an underground facility at Drakelow near Kidderminster and, by the end of the war, over 57,000 Hercules engines had been produced.

Rover's contract for aero engines continued past the end of hostilities, and so the first postwar saloons did not resume production until December 1945, although arrangements were in some disarray. Considerable time was spent in 1944-1946 developing a new baby car, known as the M-Type or M1. This was a small two-seat coupé with a 77in wheelbase and a length of just 160in, powered by a 699cc four-cylinder 28bhp engine. Looking not unlike a Fiat 500 'Mouse' of the day, up to three prototypes were built that used an aluminium chassis. Initial plans were to produce 15,000 full-size P2 saloons and 5000 M-Types. However, plans for the M-Type were abandoned in late 1946 when it was realised that it might not have sufficient sales appeal, and this left a massive gap in Rover's production plans.

The company had only managed to acquire a government licence to produce a maximum of 1100 cars with strict rationing of steel, particularly strip steel for bodies. By May 1947, the Lode Lane factory was reduced to working just four days a week, alternate weeks only. Things were looking dire for Rover's survival.

Maurice Wilks had a holiday property on Anglesey, called Tros-yr-Afon. Like many landowners he acquired a war-surplus Willys Jeep to use up in Anglesey in 1945 and soon came to appreciate the unique qualities that it offered as a small agricultural runaround vehicle, and had proved its worth as a 4x4 during the harsh winter of 1946-47. It could

The Wilks family

The Wilks family were part of the Coventry automotive aristocracy in the early part of the 20th century. They were the principal name behind Rover, with several members of the family being involved in the company from 1929 right up until the mid-1980s. To avoid confusion, here is a brief summary of the principal players.

Spencer Bernau Wilks was born in May 1891 in Rickmansworth. Having served as a Captain in the First World War, he joined the motor industry with Hillman in 1918 and went on to marry one of William Hillman's six daughters. Having married into the family, in 1921 he was made joint Managing Director of Hillman, together with brother-in-law John Black, until Hillman was bought out by the Rootes brothers in 1928. Black later went on to become Managing Director of Standard-Triumph up to the 1950s.

Spencer subsequently joined the Rover Company in 1929 as Works Director, becoming Managing Director from 1933, and by the mid-1930s he was living in some style at Street Ashton House, near Rugby, where he brought up his three children.

In his latter days he was Chairman of Rover from 1957-62, when he formally retired and moved to Islay, Scotland. At this point another brother-in-law, William Martin-Hurst took over as Chairman. Spencer accepted a nominal position as President of Rover, only finally resigning from the Board in May 1967. He died in April 1971.

Spencer Wilks. (Courtesy BMIHT)

Maurice Ferdinand Carey Wilks was some 13 years younger than his brother Spencer. Born in 1904, he spent two years in Detroit with GM from 1926-28 as part of his engineering training. On returning to the UK he too joined Hillman as a planning engineer, then joined his brother at Rover as Chief Engineer from 1930.

Maurice was the key member of the Wilks family during its heyday in the 1950s. He was not only the Chief Engineer, but also the main stylist until the mid-1950s, and was the instigator of the Land Rover. In 1956 Maurice was promoted to Technical Director, with Robert Boyle taking over his role as Chief Engineer.

He lived at Blackdown Manor, Leamington and also had the holiday property called Tros-yr-Afon on Anglesey. In September 1963 Maurice Wilks died aged just 59 at his home in Anglesey, and is buried in the churchyard near there.

The next generation of Wilks became involved from the late 1940s. Nick Wilks was the son of Spencer and joined Rover, working for some years in Engineering. His cousins Peter Wilks and Spencer King became more prominent in the company, and as car-mad boys they were all exposed to the company and its fascinating developments in cars and jet engines from an early age.

Maurice Wilks. (Courtesy BMIHT)

Peter Michael Wilks, son of Geoffrey Wilks, was born in March 1920. Leaving school at 17, he was apprenticed to the machine tool industry from 1937-40, then joined the RAF for the duration of the war. Itching to enter the company, he became a service engineer at Rover from 1946-50. Peter also built a Rover Special for club racing in 1948, before setting up the Marauder Car Company with colleagues George Mackie, Jack Gethin and his cousin 'Spen' King. Following the demise of Marauder, he became service manager at JW Gethin Ltd from 1952-54, before rejoining the family firm as production manager.

From 1954-56 he was general manager of Rover Gas Turbines, after which he held various posts in Engineering and was heavily involved in the conception of the P6 saloon. Following the untimely death of his uncle Maurice, Peter became Technical Director on 1 January 1964, and successfully led the company through the 1960s. He retired prematurely in June 1971, and sadly died just over a year later.

Spencer 'Spen' King, born in 1925, is often regarded as 'father of the Range Rover'. After leaving school in 1942, he was first apprenticed to Rolls-Royce. Joining Rover in 1945, he became involved in gas-turbine development, before leaving to set up the Marauder Car Company with cousin Peter Wilks. Spen then rejoined Rover and was made head of New Vehicle Projects in 1959.

He was generally known as 'Spen' around Rover, not least to avoid confusion with his uncle. He was a hugely gifted engineer who went on to head up Rover-Triumph engineering under British Leyland, and, later, BL Technology based at Gaydon. Spen King retired in 1985 and died in June 2010 after suffering complications following a cycling accident.

Early days of Land Rover design

Maurice Wilks (left) talking at the drawing board with Robert Boyle.

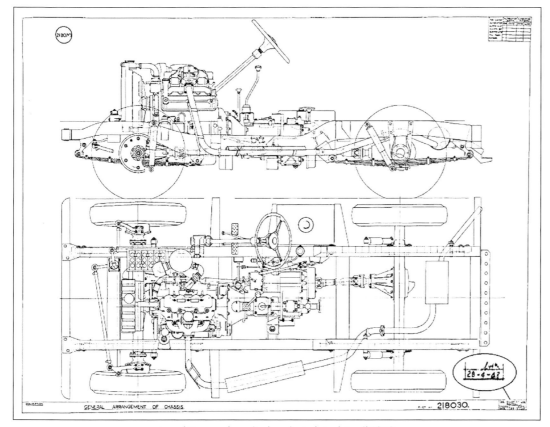

Land Rover chassis drawing, dated April 1948.

do jobs on the farm such as pulling a plough, pulling a harrow or collecting and chopping logs using the power take-off, and Maurice started to outline his thoughts for a similar vehicle that could double as a light tractor and off-roader. Inviting his brother up one weekend in April 1947, he drew the shape in the sand on Llanddona Beach in Red Wharf Bay, the outline of which we now recognise as the Defender.

The two brothers quickly realised that producing a similar light truck to the Jeep as a temporary measure to keep the factory going could be their salvation. At first they considered producing something along the lines of a half-track Ford with a big side-valve V8, but abandoned this idea in favour of a lighter vehicle.

Design work commenced straight away. To save time, two further Jeeps were bought. The first prototype (known as J-Model) used one of the Jeep chassis with a Rover gearbox, rear axle and a small Rover engine. A second prototype was made that summer using the other Jeep chassis frame, with a centre steer layout, a two-range transfer box, and 49in axles. This was fitted with an experimental Rover P3 engine. However, trials soon showed that the splayed legs driving position and awkward access were not worth pursuing, even if the aim to replicate a tractor for tax purposes was sound.

The bodies were very rudimentary, being an open Jeep-style without any doors. The first body was simply a wooden construction, but was later rebuilt with an aluminium tub and drop-down tailboard. The windscreen could be folded flat and held down with spring catches, as on the Jeep.

By September 1947, Maurice Wilks reported his progress to the board, which had considered the position and decided that "an all-purpose vehicle on the lines of the Willys-Overland Jeep was the most desirable. Considerable research had been carried out on this vehicle by our research department." Thus it was agreed to sanction the further development of the vehicle with a small team headed by Robert Boyle

Land Rover – 70 years of success

Prototype in Lode Lane workshops 1947. Note the very rounded front fender. Initial aluminium body tub was very rudimentary. The prototype was scrapped by 1949.

Period shot of the centre steer prototype, 1947. It is believed only one of the seven prototypes had the centre steer layout. (Courtesy BMIHT)

Early days of Land Rover design

A recreation of the centre steer prototype has been built, and is now owned by the Dunsfold Collection. (Author's collection)

Close-up of the centre steer layout. With its restricted access and awkward driving position that straddles the bell housing, it is easy to appreciate why the centre steer layout was abandoned. (Author's collection)

Land Rover – 70 years of success

and Arthur Goddard under the direction of Maurice. A team of five section leaders from the drawing office was assembled, comprising Gordon Bashford (chassis), Tom Barton, Frank Shaw (both transmission), Joe Drinkwater (engine), and Sam Ostler (body engineering).

Following these two prototypes, an initial sanction of 25 pilot vehicles was ordered; later increased to 50, which were largely completed by early 1948. These pilot build vehicles were assembled in the Experimental Workshop: all slightly different specifications to test out various configurations as the design progressed.

To develop the production vehicle in just seven months, the team utilised existing components wherever possible. The new 1595cc petrol 50bhp engine from the forthcoming Rover P3 saloon was used, axles derived from the Rover P2 car with a 50in wide track. A standard Rover 4-speed gearbox was employed with a new transfer box to take the drive to the front wheels, with permanent 4WD and a free-wheel device to the front wheels for sustained road use. The two ranges gave four speeds for road use and another four speeds for hauling off-road. As part of the initial design brief to make it something that could replace a tractor, the engineers were instructed to include a rear Power Take-Off (PTO) for agricultural implements and winches for tree root cutting.

Little capital was to be spent, as they couldn't wait for tooling to be made and the production run was envisaged to be short. For simplicity, the Rover-designed chassis kept the same 80in wheelbase as the Jeep, with similar approach and departure angles. Rather than a C-section forging, the chassis was designed to be made from flat steel offcuts, without tooling. The side sections

Pilot production vehicle build, 1948. Fifty pilot vehicles were ordered, but only 48 were built. Note how this is a fenced-off area within the main South Block assembly hall, with production lines still being installed. (Courtesy BMIHT)

HUE 166 was the first pilot build vehicle, R.01, released to the despatch department 11 March 1948. It was reacquired by Land Rover from a Warwickshire farmer in 1967, and has now been restored in the original Light Green, with khaki-coloured canvas tilt. (Courtesy BMIHT)

Early days of Land Rover design

From June 1949 Land Rovers were painted Bronze Green. Seen here is an early example – note the optional trafficators fitted high up on the A-pillars. After May 1950, the wire mesh grille no longer covered the headlamps.

Three-abreast seating added to the appeal of the Land Rover. In 1951, sidelights were moved from the bulkhead and mounted on the front fenders. This is an early brochure from 1952, when square-backed seats replaced the earlier spoon-back style.

were cut out with the necessary curved profiles, and the top and bottom plates were simply seam-welded at the four corners to produce a very strong closed box-section chassis.

A common Land Rover myth is that the bodies were built from aluminium because of the steel restrictions after the war, which was partly true. Speaking to *Classic & Sports Car* in 2017, Land Rover expert Roger Crathorne offered another explanation: "Maurice Wilks didn't like the way that his ex-WW2 Jeep rusted thanks to the steel body. Aluminium provided an obvious benefit, but Wilk's neighbour also owned an aluminium works so that probably had something to do with it!"

This aluminium works was Birmetals Ltd in Quinton, Birmingham, part of the Birmal Qualcast Group. It had pioneered a new aluminium alloy in 1929, trademarked 'Birmabright,' which was used in lightweight sheet form for aircraft production during the war. The aluminium alloys were provided in various types from 1 per cent to 7 per cent magnesium plus a tiny manganese content with different temper conditions, and were designed to be work hardened by cold forming into shape. The body was designed to be simply assembled with minimal tooling, using sheet steel body gussets and brackets.[1]

The Land Rover was launched at the Amsterdam Motor Show on 30 April 1948, with the P3 also being introduced that spring. The £450 launch price was kept down by making the doors, canvas tilt, sidescreens, passenger seat, spare wheel carrier and spare tyre as

---

1. The body of the original Land Rover used Birmabright BB2, with 2 per cent magnesium – a slightly thicker gauge than later Series models. Starting with the Series IIA, Land Rover switched to BB3 (3 per cent magnesium) and panel thickness decreased. Birmetals closed its factory in 1980 after losing money in 1977-1979.

Land Rover – 70 years of success

extras. However, these were standardised later that year, when the price was increased to £540.

The Land Rover was in full production by July 1948, and sales soon took off for what was a unique vehicle, planned to be built at the rate of around 50 per week. Much to its relief, Rover soon realised it had a real winner with the Land Rover – 8000 were made in 1949, and 16,000 in 1950. It was not only profitable for the Rover Company, but produced valuable foreign currency earnings for the UK, soon being sold in over 70 countries. Production quickly reached 500 per week – far more than the all-new P4 saloon, introduced in autumn 1949.

The first derivative was launched in that first year, too. One pilot build vehicle was supplied to Salmons-Tickford for development of a Station Wagon version. This featured two sideways-facing rear seats on each side in green leather, giving a seven-seat layout, with the rear seats accessed by a two-piece tailgate – presaging that of the Range Rover. First shown at the London Commercial Motor Show in October 1948, only 641 were made, mainly due to its high price of £959 with purchase tax.

The initial aims of the Land Rover as an agricultural vehicle for use on farms was aided by a friendship between Maurice Wilks and Harry Ferguson, the engineer who had developed the revolutionary tractor design employing a hydraulic three-point linkage for agricultural implements. The two men shared a common vision of finding a way to modernise British farming methods in the postwar years, with increased mechanisation as a key way to raise food output and productivity on farms from their doldrums in the 1930s.

Only one colour was offered on the Land Rover – Light Green. Myths abound that the colour was supposedly selected as it was used in fighter aircraft cockpits and was readily available as an aluminium paint colour. Land Rover expert Roger Crathorne offers another explanation: "Apparently it was down to Spencer Wilks' wife. Often the wives would be brought out to see a new vehicle and give their opinion." Her idea was to choose a colour that was similar to that used on agricultural buildings in the wartime 'Dig for Victory' campaign. It was also used on the latest Dutch barns with corrugated iron roofs being introduced on postwar farms. She suggested they use a similar colour with the Land Rover as a way of "uplifting the nation." For expediency, the production colour chosen was a standard paint colour developed for the new P3 model.

Developing the Land Rover range

Once the Land Rover had been launched and the initial engineering drawings completed, the drawing office section leaders returned to work on mainstream Rover car projects. However, Robert Boyle, Arthur Goddard and Tom Barton were retained as the core Land Rover team to further develop the vehicle, and the results of their efforts were released in September 1953 as the 86in wheelbase model.

This was the first real rethink of the ideal requirements since the hurried development based around Willys dimensions back in 1947, and developed the Land Rover into a range of vehicles – rather than a single model – for the first time. The basic vehicle was a much improved package, with 6in (150mm) added behind the seats, plus an extra 3in (75mm) rear overhang to give a considerable increase in load space volume, even if the payload was still rated at 1000lb (454kg). Pedals were repositioned and a more comprehensive instrument panel was designed with large instruments in the centre of the dashboard.

Boyle and Goddard also came up with a significant new derivative, an LWB 107in version with longer 6-foot load bed (1836mm), known

---

Land-Rover or Land Rover?

In early memos from 1947, Maurice Wilks referred to his new project as a 'Land-Rover' – with a hyphen. This name stuck, and was used as the official name of the vehicle from its launch in 1948. This spelling with a hyphen was used in all company brochures and publications up to 1978, when the company was hived off as a separate division within British Leyland. From this point, the hyphen was dropped, although usage of the name with a hyphen continued in a somewhat-inconsistent manner.

Early Land Rover 80in models sported a yellow badge. Roger Crathorne: "There was a chap called Lawrence Watts who worked in the body shop. He used to sign his name 'Larry Watts' – across two lines and with a zig-zag between the two. Maurice Wilks spotted that and said they wanted the badge just like it but with 'Land' and 'Rover' instead." (Author's collection)

Series III grille badges still sported the name 'Land-Rover' up to 1985, although the Land Rover One Ten was badged without a hyphen. The Stage 1 V8 launched in 1979 used a new grille badge, but still with a hyphen, as did the 'Land-Rover V8' side decals.

For consistency, we have used the term 'Land Rover' throughout the book.

The Land Rover logo in its elliptical background underwent several variations throughout the 1950s. The outline for the original badge was not a true ellipse and the story goes it was traced around a leftover lunchbox sardine tin that had left a greasy mark on the drawing! Most often used as white script on a black ellipse, some early vehicles employed a yellow or red script, and the logo could also be found in company literature on a yellow or green ellipse.

From its launch in 1970, the Range Rover name was always used with no hyphen.

Series III grille badge.

Early days of Land Rover design

The Tickford 80in Land Rover Station Wagon featured an ash-framed coachbuilt body with a spun aluminium spare tyre cover on the bonnet. It was assembled by Tickford in its Newport Pagnell works. (Courtesy BMIHT)

as the Pick Up. This had a 21in (534mm) longer body with higher rear sides to the rear tray, and answered the need for a greater 1500lb (680kg) payload that many customers were now requesting. A De Luxe version was also offered, featuring thicker seats trimmed in a blue gingham pattern and door pockets. For the first time, an aluminium hard top became available for both versions, painted in a limestone cream colour.

A new Station Wagon body was announced for the 86in, with a seven-seat capacity, and introduced three signature Land Rover design features. The first was a curved 'Alpine Light' in the roof of the Station Wagon and Hard Top to allow more light into the rear cabin. The second was the tropical roof, a second aluminium skin on the roof with three ribbed reinforcing sections riveted on. The small air gap between the two skins reduced the amount of heat coming into the cabin and allowed air to enter via a hidden interior roof vent. The accompanying final feature was the addition of opening vents under

Land Rover – 70 years of success

Early days of Land Rover design

Opposite, clockwise from bottom left: Post-1952 80in with external door handles; The 86in had a redesigned bulkhead with a higher scuttle and more rake to the front screen. It also introduced the recessed door handle feature. Note the two rivets in front of the rear wheel – a recognition point for the 86in; Comparison of the rear tray load space between an 80in and 86in model; 80in cabin with the early style of instrument panel and floor-mounted pedals. Blumels spring spoke steering wheels were fitted up to the 1960s. (Author's collection)

the front screen as a standard feature, to further aid airflow into the cabin, although the fold-down screen feature was still retained.

On its introduction the 86in was available only in light grey with blue PVC upholstery, but in 1954 grey and blue paint colours were made available across the 107in range, in addition to Bronze Green. Chassis were painted blue or grey, too, with wheels painted the same as the chassis colour, although after a couple of years chassis were all painted black.

In 1956, the ten-seater 107in Station Wagon was introduced. This had a particularly homespun appearance to it, with ugly galvanised cappings surrounding the rear door apertures. The model was rather short-lived, as one year later the Land Rover was revised yet again to accommodate the fitting of the forthcoming new 2-litre diesel engine, introduced some months later. At this stage the wheelbase

86in Station Wagon. Note the tropical roof panel and long Alpine roof light. (Author's collection)

Land Rover – 70 years of success

The 107in Station Wagon had a very homespun appearance. (Author's collection)

was increased by a further 2in (50mm), achieved by moving the front spring hangers forwards on the chassis. The wheelbase was thus standardised into the familiar 88in and 109in versions for the subsequent Series Land Rover models, with the two versions referred to in sales terms as Regular or Long.

That same year, the law lords judged the Land Rover to be a 'car-type' vehicle, which meant it was no longer restricted to 30mph maximum speeds on the road as a commercial vehicle. This had followed some years of dispute about the exact classification of the Land Rover, resulting in local prosecutions by police if drivers were caught speeding when the vehicle had been purchased as a commercial vehicle without payment of purchase tax. Another issue was the use of tax-free red-dyed fuel, as supplied to farms. This was fine for agricultural vehicles but not for road vehicles and farmers could be prosecuted if found with red fuel in their vehicles and driving at more than 30mph on the road.

January 1956 saw the first expansion by the Rover Company to the north of the Lode Lane entrance, with the construction of

Early days of Land Rover design

A pair of Royal Review vehicles. The right hand one is an 86in model, the left is an early Series II with a special grille design.

Tom Barton in 1959. (Courtesy BMIHT)

a 100,000sq ft dispatch area. Rover was suffering the first of several waves of factory space difficulties, and started to acquire various smaller sites around Birmingham as a way of alleviating these bottlenecks. The Perry Bar plant was purchased in 1952, and Percy Road in 1954 to add capacity to the gearbox machining and assembly that was carried out at the existing Tyseley and Acocks Green plants. Later on, Tyburn Road was added in 1964 to make transmissions and chassis, Garrison Road in 1965, followed by Tyseley No 2 plant in 1969.

The arrival of David Bache
Up to this point the company had relied on the instincts of Maurice Wilks as Engineering Director to style the cars, including the P4 saloon. Although untrained as a designer, Maurice had evolved a conservative yet finely balanced look for Rover cars, but he realised he had reached his limits in terms of progressing Rover's styling and needed a more talented designer to take on the task. Thus it was that David Bache joined in late 1953, and to appoint a young 28-year-old as Rover's first stylist was seen as a very brave step for the family-run business.

Bache inherited a small department of four skilled modellers to interpret his designs, and his first job was to update the P4 saloon. For the October 1954 face-lift he introduced a raised trunk line and wider rear window, divided into three pieces, no doubt because Maurice Wilks had concerns about the strength of a large one-piece curved glass. Bache also designed new vertical rear lamps incorporating flashing trafficators to meet the latest UK lighting regulations, and revised the front fenders with sidelights mounted in chromed pods atop the fender.

In October 1954, Bache was sent to visit the Paris Motor Show, where the new Facel Vega FV and Ghia Chrysler GS-1 Special show car made a big impression on him. On his return he was asked to start work on a new larger saloon for Rover – the P5 – and the initial clay models were highly influenced by the Italian coach-builder Ghia's designs for Chrysler, together with the French Facel Vega.

21

Land Rover – 70 years of success

Early days of Land Rover design

The Farina drophead prototype 1952, based on the P4. (James Taylor)

David Bache. (Courtesy BMIHT)

There were other influences closer to home, too. In late 1952, Farina in Italy had been commissioned to produce a coupé and drophead convertible on the Rover P4 chassis, prior to Bache's arrival. The Farina Coupé served as a starting point for Maurice Wilks' ideas for the P5, but he realised he needed a full time designer to bring this project to fruition, and was too far out of his artistic depth where modern styling was concerned, thus reinforcing his decision to seek outside help.

A further project was the Rover T3 gas turbine car for which Bache was asked to provide styling input, and it was through this project that Bache discovered his first design recruit in spring 1956. Tony Poole was a fitter in the gas turbine department, having been involved with jet turbines since his time in the RAF during the latter days of WW2. Poole sketched suggestions for badges and other details on the T3, and on the strength of this found himself nominated for a permanent job in the styling studio as Bache's assistant.

Land Rover – 70 years of success

Rover styling studio in the 1950s

Maurice Wilks was not only Chief Engineer, but had styled every Rover since the Pilot 14 of 1932. He had developed a good eye for design, and worked closely with his experimental body engineer, Harry Loker, to develop a handsome range of four-light and six-light saloons. Prewar body construction was steel panels on ash frames, and new body prototypes could be knocked out fairly quickly using similar methods.

After the war, Wilks and Loker were unusual in adopting Plasticine modelling clay rather than wood to develop full-size models for their P3 saloon studies. Plasticine was also used for early scale models of the P4 saloon, which used the 1947 Studebaker as a template for its design. The only other designer using clay in the UK at this point was Walter Belgrove at Triumph, and it is probable that John Black (Maurice's brother-in-law and Managing Director of Standard-Triumph) alerted him to this new material.

Frank Underwood was employed as a model maker, chiefly producing the beautifully-detailed scale models used for initial evaluations. Models were photographed in a scale diorama setting, and large scale photographic prints would be used to evaluate the design by management.

After the arrival of David Bache, the latest modelling methods as used at Austin were introduced, including the use of industrial American modelling clays, such as those made by Wilkins-Campbell. Ken Barton became modelling manager, and a handful of modellers were employed to work with this new material, particularly for the P6 programme. In the late 1950s a new engineering block was built, which allowed Bache to relocate to a small styling studio housed on the first floor of the South Works, facing the new engineering block.

Bache fitted in well with the patrician management style of Rover, who appreciated his assured balance of line and sense of form, which resulted in the P5. Upon its launch in October 1958, the P5 was regarded as a fine alternative for those who found a Daimler too staid and a Jaguar too flashy. Bache's design fitted the bill perfectly, with its muscular form influenced by the Ghia Chrysler showcars, and, in particular, the Chrysler 300. The P5 was also planned to include a 'hard top' or 'sports saloon' derivative, although this four-door coupé version was not launched until 1962. In this form the high-waisted Bache design was even more arresting, and has gone on to become the most highly sought-after version of the P5 amongst classic car owners.

Maurice would still exert a strong influence over Bache, however. In his early days his sketches were rejected by Wilks as being too daring because "The Rover Company does not make head-turners!" Indeed, this view was epitomised by the nickname of 'Auntie' that the P4 saloons earned in the 1950s, as they represented the same staid middle class respectability as the 'Auntie' BBC.

Talking to author Graham Robson about Maurice's continued influence in the studio, Bache commented "Not that he didn't get in on the act. But he used to do it in the most diplomatic way. He'd ask rather tentatively whether *this* line couldn't be moved to *there*, and I would tell him why not, but in the end we would usually compromise."

Bache also recalled that when the Wilks brothers periodically came down to the styling studio to view progress on projects, they were provided with a special pair of P5 seats, known as the 'Wilks Thrones.' "They would rub their hands, look gleefully around, and ask each other 'Is it time to kill off Auntie yet?'"

Modellers prepare an early quarter-scale P6 clay model (Courtesy BMIHT)

Series II Land Rover

Although the styling department was busy on the P5 saloon, another project was pushed its way during 1957. This was the redesign for the Series II Land Rover, to be introduced at the Amsterdam Motor Show in April 1958. In 1956, Maurice Wilks had been promoted to Technical Director, with Robert Boyle taking his role as Chief Engineer for Rover cars and Tom Barton becoming Assistant Chief Engineer for Land Rover. The revised vehicle had been drafted out by Boyle's engineering team with wider tracks front and rear that demanded a broader body, and the opportunity was taken to involve Bache's team, the first time that Styling had been involved with the Land Rover.

David Bache and Tony Poole were tasked with tidying up the body, and both would have relished the opportunity to revisit the crude appearance of the Land Rover and inject it with a stronger symbolism befitting a Rover product.

"The Land Rover [as we found it] was so obviously 'right' already that we couldn't just dash in with some obvious improvements" commented Bache to author Graham Robson in 1979. "It just wasn't the sort of machine that cried out for decoration, and because of the nature of its work there was no point in putting a lot of delicate shape into the panels."

As a result the upper cabin width was left largely unchanged, but the opportunity was taken to introduce a small amount of tumblehome in the side panels to alleviate the awkward flat sides of earlier models. A simple rolled shoulder line generated from the wider front fender ran through the doors and into the rear fender, giving a more robust look to the vehicle, with the twin round Lucas rear lamps nestling in the rear corner capping. A new 'De Luxe' bonnet pressing was designed with a rolled front edge that implied a more solid construction, although the simple blade-style was retained for use on basic 88in models.

Door upper frames were simplified and painted in body colour, and the number of galvanised cappings and brackets was minimised to clean up the appearance, particularly for the Station Wagons. Glass replaced Perspex in the sliding side windows and cast iron door hinges were introduced, also painted in body colour. Bache and Poole also developed a range of five standard body colours plus Red, Sand and Dark Grey as options on the home market to add to the appeal.

Sill valances under the doors tidied up the side view and hid the exposed chassis and exhaust that were visible on earlier models. The fuel filler was relocated in the offside rear fender behind the door, making it much easier to refuel without having to lift the driver's seat. At the front, the mesh grille was modernised and a new valance added beneath to tie the volumes of the front end together visually. New lighting regulations demanded the fitment of orange trafficators which were paired up with the sidelights horizontally along the top of each front fender and indexed with the square front licence plate usually fitted in the UK, with a matching square plate fitted to the offside rear panel. Finally, a much airier truck cab was designed with curved rear corner windows.

David Bache with first models for the P5, summer 1955. (Left to right) Harry Mills, Dennis Lyons, Bache, unknown (modeller?), Frank Underwood. Note the fine metal detailing of the painted wood models, which are on a scale diorama tabletop. The location is Street Ashton House, near Rugby, family home of Spencer Wilks. (Courtesy Mark Bache)

One of the reasons to update the Land Rover to Series II at this time was as a response to emerging competition in the 4x4 market. Austin was known to be developing the Gipsy 4x4, while Fiat had recently launched the first Campagnola model. More seriously, Willys was casting around for a European partner to produce the Jeep, and there were rumours that Standard-Triumph in Coventry had been in discussions with them. While Jeep sales had been largely confined to North America, life was beginning to get a bit tougher for Rover and, while it had managed a great run so far, it could not afford to become complacent about the continued success of the Land Rover.

In fact, later on in 1958 there were discussions between Rover and Willys about collaborating on a modified 83in Land Rover. The body would be the CJ-5 Jeep with 83in wheelbase, but the chassis was Rover, as were the engine, bulkhead, instruments and other components. A certain Colonel Jack Pogmore had been drafted in within Land Rover for a senior position above Barton to strengthen links to the MoD, and he was tasked with leading the discussions with Willys. In the board minutes of 7 July, it was reported that "The 83 Land-Rover-Willys-Jeep has been completed and is available for assessment. This vehicle requires remarkably few major modifications." For whatever reasons, both sides soon abandoned the idea.

There were also talks with Standard-Triumph about a possible merger at this time, and some exchange of company information that each was interested in the 2-litre car market, which would result in the Triumph and Rover 2000 models, both launched in 1963.

The company was wise in updating the model at this point, as it reinforced the Land Rover's position as the leading 4x4 in the market. It was also smart in not wasting money, and sought ways to minimise the design process. Allegedly, the Series II range went from the drawing board to the first metal prototype in a mere six weeks, while early P5 concepts were reviewed by management not through time-consuming full-size clays, but via cheaper photo-montages using scale models.

The Road-Rover

The idea to develop a more road-based Station Wagon concept was first discussed in 1951, possibly as a reaction to the poor

Land Rover – 70 years of success

Classic period shot of a standard 88in Series II Truck Cab, with blade-type bonnet, as launched in 1958. The bonnet-mounted spare tyre was a popular option to maximise load space in the rear.

acceptance of the Tickford Station Wagon. Maurice Wilks directed Gordon Bashford to draw up a two-wheel-drive estate car type of vehicle using a P4 chassis with a shortened wheelbase, which would be known as the 'Road-Rover.' Wilks and Harry Loker then added a simple flat-sided body so that the project could be evaluated further, and the first prototype, christened 'The Greenhouse,' was running in 1952.

The project seemed to gather interest within the company, and in April 1953 the board approved the Road-Rover as a production programme. At this point there was a consensus that the body design needed a major rethink, and Wilks and Loker started tentative schemes for a less utilitarian body, although restyling work for the P4 and ongoing Land Rover work meant it was a slow burn project for the next couple of years.

Early days of Land Rover design

Road-Rover mock-up, circa 1951. A Maurice Wilks and Harry Loker exercise. This has lower front fenders and a more curved roof than the later prototypes.

'The Greenhouse' outside the main office block at Lode Lane. Twelve prototype Road-Rovers were made. MWD 716 can still be seen at the British Motor Museum, Gaydon. (Courtesy BMIHT)

Land Rover – 70 years of success

Construction of the first prototypes of the resulting 'Road-Rover Series II' was started in 1956, and they were running by 1957. For these cars the wheelbase was increased to 97in, the front suspension was independent (like the P5) and the front brakes were now discs. According to Bache, he had nothing to do with the styling, which was a peculiar mix of Rover P5 and Chevrolet, and was probably the last attempt by Maurice Wilks and Harry Loker at body styling. According to Gordon Bashford, as related to author Graham Robson "[The management] thought the original was too austere, and asked for changes to the light alloy bodyshell, involving the shape and needing complicated pressings. In this way it got bigger, grander, heavier and more costly."

A look at the Engineering Progress Reports for 1958 reveal how the development evolved. At the meeting on 23 January 1958, it was reported that:

"Item Road-Rover: A great many design problems need to be solved before a satisfactory specification could be issued. The major problems are door locks, glazing, sealing of curved glasses and general structural strength. A final evaluation of the present specification is expected by end of February. When this is complete three major questions will have to be answered:
- Can the required design alterations be completed within acceptable terms of reference for timing, costs and production problems?
- When can scheme sheets and drawings be available, bearing in mind that other commitments will not allow DO work to start for 6-8 weeks?
- What problems will have to be faced by production as a result of the required modifications, also bearing in mind possible derivatives?"

By 13 March it was reported that "The test on Road-Rover number 7 was completed on time ie by end of February, and showed that the general basic structure is satisfactory. Many modifications known to be required were not incorporated in this vehicle and the whole drawing work on these needs to be done, and in certain cases design problems will have to be solved. Owing to other work in the Body DO this cannot be started until middle of this month. A programme has been prepared that includes the building of two further prototypes and the issue of a production specification on January 31 1959."

At the same meeting Robert Boyle reported that "P5 and P5 Hard Top will impose a considerable load on the DO in terms of possible

The Series II Road-Rover has hints of the Chevrolet Nomad, with a big Rover grille and wrap-around screen, but remains very ungainly, despite an attempt to break up the slab sides with a two-tone colour scheme.

Interestingly, it has the floating roof treatment, all-round glazing, and split rear tailgate like the later Range Rover. One rumour is that the project was cancelled once the FA Vauxhall Victor Estate was launched, offering an unbeatable estate car package for just £858.

mods for several months," and "The production specification for the Series II 88in SWB now issued. We expect to complete all drawings for 109in LWB by end of April." Clearly the drawing office was under considerable pressure to complete work on a number of projects and it is not surprising that work on the Road-Rover was increasingly seen as a hindrance.

The project continued into the summer however. On 8 May Boyle reported that "The Sales Department's report is being prepared. In the meantime the programme proceeds as previously reported ie a target production date for October 1959."

On 13 June he reports that "Design is proceeding to programme for the final production specification. Discussions are proceeding with Sales Department as to whether this should be marketed as an 'austerity' version as a 2WD extension of the Land-Rover range of vehicles with a more elaborate specification as an optional extra, or whether we should proceed with the more elaborate specification for all vehicles, as at present envisaged."

This idea to produce two versions of the Road-Rover was further discussed at the meeting on 7 July: "Design of the De Luxe version is proceeding to schedule. Following discussions at the previous meeting, a decision was made to proceed with a 'standard' model, incorporating Land-Rover parts and finish wherever possible. While it will be difficult, every effort will be made to have this available at the same time as the De Luxe model, but this will depend on the latest date when modifications from the De Luxe can be accepted by Purchase and Production."

Boyle also reported "Certain difficulties on Series II Station Wagons, which have recently become apparent are being dealt with and Mr Pogmore and Mr R Harris are collaborating on this work." Plus "P5: tests on Production Prototypes. There is a lot of work to do on these to bring them up to a satisfactory standard." Clearly Rover could not afford to use valuable engineers on the Road–Rover while these other critical programmes were at risk, and the project was cancelled soon afterwards.

In total, 11 Series II Road-Rover prototypes were produced, all built on modified P4 chassis with RWD only and torsion bar independent front suspension like the P5. Most of them were fitted with the 1997cc Rover 60 four-cylinder engine with a four-speed manual gearbox plus overdrive. The body used a steel underframe with light alloy skin panels, but the ground clearance of just 5.4in (136mm) meant it was not suited to off-road work at all.

The Road-Rover got tantalisingly close to production. During 1958 pilot build production took place with a view to series production in 1960-61, but it never happened.

To be honest, by the late 1950s its moment had passed. The idea of a utilitarian no-frills station wagon with two doors had been overtaken by the mass-produced family estate car. Cars such as the Standard Vanguard and Hillman Minx station wagons had evolved from two-door to four-door format by now, and been joined by new models such as the Vauxhall Victor FA and Austin Cambridge estates, all available at around £860. They offered everything the Road-Rover could offer and more: four doors, greater loadspace and more sophistication through use of a monocoque bodyshell designed around passenger car comfort. Not to mention far wider dealer networks for sales and service backup.

Lower down the market the two-door estate car market was well-established as a functional vehicle for families on a £650 budget, with models such as the Ford Squire, Hillman Husky and Morris Minor Traveller. Meanwhile, the latest Austin A40 offered a novel split tailgate hatchback format, with neat styling by Pininfarina – all for a modest £640. With the Road-Rover priced at around £1100, Rover could not hope to compete.

1960-1978

Chapter 2

David Bache era and Range Rover

By the early 1960s production of the Land Rover was running at 750 per week, with over 250,000 sold by November 1959 – almost twice the rate of Rover's saloon cars. This 'stopgap' offshoot had taken the company in quite another direction to that planned in the 1940s. One can imagine that, at heart, the Wilks brothers had envisaged filling production at Lode Lane with a nice range of high quality saloons, possibly a few coupés and convertibles too, much as it had done prewar.

Instead, they had ended up with this quasi-military vehicle, albeit one that was proving highly lucrative, providing security and profits for Rover. Without Land Rover, Rover itself could well have fallen into the financial difficulties that befell companies such as Triumph and Borgward in the early 1960s, with falling sales and insufficient revenues to fund future models. By contrast, the company could embark on the huge investment to produce the next car in its plans, the advanced P6 Rover 2000, which was launched in 1963.

Even so, it seemed Rover struggled to fully embrace the Land Rover, content to allow production to continue without giving it the full resources it deserved to grow into a dominant global brand for the 1960s. After the Series II, development was carried out on a piecemeal basis with the minimum of long-term investment in either engineering resources or production facilities.

To provide space for the new P6 saloon, a north block at Lode Lane was constructed to provide a body assembly area, complete with a new paint shop. The P6 was highly innovative in its engineering and design, resulting in industry awards such as the Dewar Trophy for safety in cars, and was the first model to receive another newly established award – Car Of The Year. Bache's involvement on the P6 brought him into the public realm for the first time and the publicity was instrumental in raising awareness that professional car design existed as a serious branch of industrial design. Up to that point, it tended to be engineers such as Maurice Wilks or Alec Issigonis that were credited in the media as the creators of new cars, with 'stylists' rarely getting a mention.

Opposite top: Lode Lane plant, late 1950s. The South Works forms the main block to the right of the main drive off Lode Lane. Two new blocks have been built on the left as dispatch areas, and these would be extended in 1960 to become the North Works, a new block to produce the P6 saloon. (Courtesy Roger Crathorne)

Opposite below: Early mock-up of the P6. The roof of the Lode Lane styling studio was used as a secure viewing location for secret models. The P6 cost £10.6m to develop. (Courtesy Roger Crathorne)

Land Rover Design – 70 years of success

David Bache

David Ernest Bache was born on 14 June 1925 in Mannheim, Germany. The Bache family was living in Germany because the father, Joe, was managing a football team there. Joe Bache was a well-known sportsman who had captained his Aston Villa side for 11 years, had played football for England, and also played cricket for Worcestershire.

On returning to England, David went to school in Cheltenham, followed by studies at Birmingham College of Art and Birmingham University. It was a comfortable middle class upbringing, but David was determined to enter the new profession of car design. In those days the only way into the motor industry was via an apprenticeship.

David had a brother – also called Joe – some 15 years older than him who had joined Austin in 1930 as a Sales Manager, and by 1952 was promoted to Sales Director for the company. It was therefore not surprising that young David joined Austin as an engineering apprentice in 1944 and by 1948 had joined the styling department at Longbridge, led by Ricardo (Dick) Burzi. Working in Burzi's studio he would have learned how to sketch and build wooden mock-ups, although Plasticine was sometimes used for scale models. During his time there he designed the instrument panel for Burzi's new Austin A30 saloon, which came out in 1951.

Such was the rarity of competent designers in the Midlands that he was soon headhunted by Maurice Wilks, and was invited to join Rover in late 1953. Aged 28, he was given the title Chief Styling Engineer and soon set about with his first task, a face-lift for the P4 saloon. This was followed up with the P5 Saloon in 1958, the P5 Coupé and the radical P6 Rover 2000 in 1963.

Always smartly dressed, Bache stood out amongst his engineering peers at Rover. His preference for double-breasted suits with a waistcoat, protruding cuffs, large neck tie and heavy jewellery were set off against his helmet of thick slicked-backed hair with no central parting.

Much of his sartorial elegance was gained from his wife's family, who owned a series of men's outfitters in Birmingham, and Bache and his wife Doreen inherited one of these shops on the Bristol Road, complete with a flat above where they lived after they were first married. Despite this, he always preferred to have suits made by his own tailor.

The 1976 Rover SD1 was another radical Bache design that promoted the idea of a large hatchback executive car and was widely copied in subsequent mainstream models such as the Ford Sierra and GM Cavalier. Bache headed up all of British Leyland car styling from 1975 and was responsible for revising the ADO88 into the Austin Metro, together with the Austin Maestro and the Montego designs.

After Bache was sacked by Harold Musgrove in 1981, he set up David Bache Associates, engaged primarily in product design work. He worked for Volvo and other accessory manufacturers attached to the motor industry before his death from cancer on 26 November 1994, aged 69.

David Bache was one of the few car design 'names' in the 1960s. (Courtesy BMIHT)

Bonnet Control 129in truck, 1963. This is one of the prototypes built for military trials for the Belgian army in the early 1960s. Six prototypes were built; this is one of the surviving two, now housed at the Dunsfold Collection in Surrey. (Courtesy Dunsfold Collection)

David Bache era and Range Rover

Land Rover developments and the 109in Forward Control

In the early 1960s Rover Engineering Department was split into three sections: Car Engineering, with Dick Oxley as chief engineer, Land Rover Engineering under Tom Barton and the New Vehicle Projects think-tank, led by Spen King. All three subsequently reported to Peter Wilks – appointed Technical Director on 1 January 1964 following the untimely death of his uncle Maurice – who successfully led Rover Engineering through the 1960s.

The mainstream Series II Land Rover models continued with very little alteration throughout the 1960s. The Series IIA from September 1961 introduced an enlarged diesel engine with 2286cc, and the six-cylinder petrol engine was shoehorned into the 109in LWB version from April 1967, but neither required any styling input.

A small Land Rover 'New Projects' group had been established in the late 1950s, which resulted in two models with heavier carrying capacities. The first was the 129in project, a large 30cwt 4x4 truck that could appeal to both commercial and military markets, developed under the guidance of Jack Pogmore. The initial target was the oil fields of Middle East, where the Dodge Power Wagon was popular, albeit based on a dated wartime design. The 129in used the P5 six-cylinder petrol engine, but remained underpowered and suffered severe overheating during development. The project was abandoned in 1963, and Jack Pogmore left soon after – probably much to Barton's relief.

The second project was another 30cwt design commenced in 1959 that proved more effective. This was the 109in Forward Control that used 75 per cent of existing 109in Long components, including the chassis. To achieve the Forward Control layout and 30cwt payload a new overframe was added. Project engineer Geof Miller worked on the 129in and 109in Forward Control projects, and confirms that Eastnor Castle was used as a test course for the 129in – the first time this now-famous testing ground was used by Land Rover.

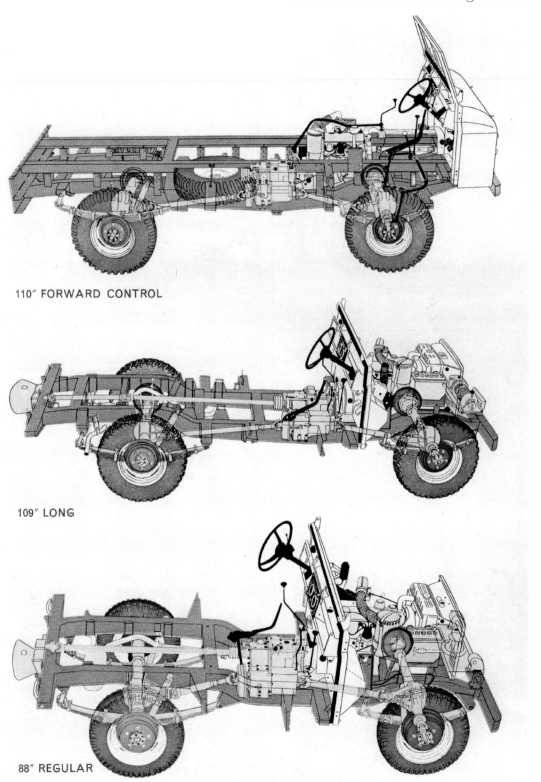

The family of chassis used in the 1960s. It is easy to see how the 109/110in Forward Control chassis was derived from the normal control 109in version, with an added overframe.

109in Forward Control was introduced from September 1962. Shown here is the redesigned 110in version with 4in wider tracks that superseded it from September 1966.

The Forward Control models carried on until 1972, when they were augmented by a much more specialised vehicle – the One Tonne Forward Control. The British Army was finding its 109in and 110in Forward Control Land Rovers were underpowered and increasingly overworked. There was a new requirement for a larger 4x4 truck that could carry a one tonne payload behind the seats and could cope with towing a 4000lb field gun.

Therefore, in 1967, Special Projects set about designing a completely new military Land Rover with a 101in wheelbase and Forward Control cab, mated to a detuned version of the Range Rover V8 engine and drivetrain. The vehicle also needed to weigh less than 3500lb so it could be airlifted by Wessex or Puma helicopters, and to meet this requirement the bodywork was designed to be quickly detached. The final feature was a rear power take-off combined with a tow hook to be able to tow a powered gun carriage, such as that developed by Rubery Owen in a 6x6 configuration.

However, the forays into larger Land Rovers and Forward Control models were proving disappointing on several fronts, and the Rover directors spent some time in the mid-1960s pondering how they might add an additional Land Rover product to boost sales. One direction was to develop a smaller vehicle and a Haflinger-type light vehicle was considered around 1964, as was a new 80in model, but the conclusion was that they would not be much cheaper to produce so profits would be slim. BMC's Mini Moke came on the market in 1964 withlacklustre sales, further dampening the enthusiasm at Solihull for a smaller Land Rover.

The full line-up of standard Land Rovers in the 1960s comprised 38 different versions.

88" REGULAR LAND-ROVER
1. Full length canvas hood
2. Full length canvas hood with side windows for export
3. Hardtop with tailboard and top hinged flap
4. Hardtop with side hinged rear door
5. Hardtop with fixed side windows (export only) tailboard and top hinged flap
6. Hardtop with fixed side windows (export only) and side hinged rear door
7. Hardtop with sliding side windows (export only) tailboard and top hinged flap
8. Hardtop with sliding side windows (export only) and side hinged rear door
9. Cab, ¾ canvas hood with side windows for export
10. Cab, ¾ canvas hood
11. Cab, open rear body
12. Chassis with cab and cab base
13. Chassis with wings, dash and seat-base

109" LONG LAND-ROVER
14. Cab and open rear body
15. Cab and ¾ canvas hood
16. Cab and ¾ canvas hood with side windows for export
17. Hardtop with tailboard and top hinged flap
18. Hardtop with side hinged door
19. Hardtop with tailboard and top hinged flap and fixed side windows for export
20. Hardtop with side hinged rear door and fixed side windows for export
21. Full length canvas hood with side windows for export
22. Full length canvas hood
23. Chassis with cab and cab base
24. Chassis with wings, dash and seat-base

STATION WAGONS
25. Station wagon 7 seater
26. Station wagon 10 seater
27. Station wagon 12 seater

110" FORWARD CONTROL LAND-ROVER
28. Cab and fixed side rear body
29. Cab and dropside rear body
30. Cab and fixed side rear body, ¾ canvas hood
31. Cab and dropside rear body, ¾ canvas hood
32. Cab and fixed side rear body, ¾ canvas hood with side windows for export
33. Cab and drop side rear body, ¾ canvas hood with side windows for export
34. Cab and platform rear body
35. Chassis and cab with subframe*
36. Chassis and cab, no subframe*
37. Chassis with wings, dash, seat-base with subframe*
38. Chassis with wings, dash, seat-base*

NOTE
Reference should be made to the Sales Department before any consideration is given to the use of chassis marked * above since certain limitations in their use may apply.

Land Rover Design – 70 years of success

Half Ton lightweight first prototype, R6796-2, in 1965 and driven by Ken Twist, the experimental shop foreman. Note the decent styling effort on the front grille, later simplified for production versions.

Military Half Ton lightweight Land Rover, introduced from 1966. This is a 1983 version.

First shown in 1972, the 101in One Tonne Forward Control was produced on a special line at Lode Lane from 1975-78, at the rate of around 20 per week. (Author's collection)

David Bache era and Range Rover

Cuthbertson tracked vehicle conversion, 1960.

The Forest Rover was developed by Roadless Traction Ltd in 1964. The wider axles shod with 10x28in tractor tyres allowed it to traverse logs and deep ditches for forestry work.

Land Rover Design – 70 years of success

Technical Sales and Special Projects

Initial discussions for a Special Projects department for Land Rover were started as early as 1952, resulting from discussions between George Mackie and Peter Wilks, but they decided there were not enough vehicles around to warrant it. Initially, the Service Department handled special requests from Land Rover customers but they were not truly equipped to handle it.

By the late 1950s the Land Rover's versatility had really gripped the public's imagination, with many enquiries coming in for conversions and adaptations. Thus, from January 1957 a Technical Sales Department was set up within Sales, headed by Mackie, with three engineers and one secretary. The main aim was to cover liaison between Rover and the specialist outside manufacturers and suppliers, and Mackie set up a proprietary approval scheme based on that used by tractor manufacturers, to ensure performance and fitness for purpose was adhered to. Retention of the Land Rover warranty was seen as a key advantage for their work. Mackie's brief from Spencer Wilks was to ensure 200 extra vehicles per year – if that was achieved then the group was worth it.

Ambulance conversion under construction.

The idea of using a Land Rover as an alternative to a tractor had not taken off and the vehicle was being bought more by public utility companies. To target this market, a new type of tailboard-mounted compressor by Bullows allowed quick adaptation of the vehicle and the idea was extended to water pumps, vacuum pumps and generators. Of course, farmers did buy the Land Rover but used it for towing horse boxes and trailers to move livestock around the farm, as well as personal transport. Real tillage work remained the preserve of tractors, and a two-furrow plough was the Land Rover's limit on medium soil, but the cost of buying and fitting a three-point linkage was too high.

Capstan winches were another early project to be tackled but then Mackie's group looked at crop sprayers and fertiliser spreading equipment as a line of investigation and the Commercial Vehicle Show in 1958 was used to showcase their initial range of approved implements. Approved suppliers included Pilchers (ambulances), Dixon-Bate (towing gear), Dowty's (pumps) and Evers & Wall (crop sprayers), while Carmichael and Sons of Worcester were engaged for fire appliances.

A sample of special conversions from the 1970s. Simon hydraulic cherry picker platforms were popular for use by electricity boards.

By the early 1960s the number of requests from Mackie's group to Engineering was starting to interfere with regular drawing office work, and Maurice Wilks decided a new arrangement was now needed. Engineering was a bit resentful of an engineering group outside its control, and so it became the Special Projects department within Engineering.

In 1965, maintenance of the 100 acres of farmland surrounding the Lode Lane plant became the responsibility of Special Projects, and it used the land to test grass cutting equipment, with a base at Foredrove Farm on site. By now Tom Barton was Chief Engineer for Land Rover, and he retained a dedicated military development section for MoD work – with Special Projects never involved at this stage. However, once Alvis was on board, the focus shifted more to military vehicles and dealing with the MoD, and the Sottorn-powered trailer project that accompanied the 101in Forward Control model was given to Mackie's team to develop.

Into the 1970s Special Projects shifted to Range Rover work, especially ambulance conversions. At first a 10in chassis extension was tried, later a 35in extension plus a 10in added rear overhang. Other ambulances used the Carmichael 6x4 chassis.

In the 1980s the group was reorganised again. Special Projects, Military Contracts, Engineering Liaison and Spanish Santana manufacturing were grouped together as 'Special Operations', under a Chief Engineer. This subsequently became SVO in July 1985, with a more focused remit as a bespoke tailoring unit of the plant, directed by Roland Maturi. Luxury conversions, specialised military vehicles or custom built workhorses could be ordered through the normal sales channels. An SVO project engineer would then draw up a detailed design specification, with vehicles built either within SVO or at an approved body builder. The new 127in model was a particularly useful chassis for conversions and was initially built within SVO. A One Ten chassis was cut in two, the extra 17in was inserted in a special jig, and everything was welded back together. After 1987, 127s were produced on the main production line.

Land-Rover Special Conversions

The cross-country performance capabilities of the Land-Rover, its load carrying capacity and availability in chassis form all contribute to the demand for this vehicle as the basis for special conversions and for the fitting of specialist appliances. We show just a few examples here. Left. Police, Fire and Ambulance service vehicles. Above. A hydraulic platform. Right. A self-contained safari-type motor caravan. A snow-blade. A crop or verge sprayer.

Land Rover Design – 70 years of success

The styling team expands

The styling team grew slowly into the 1960s as Rover product plans expanded. One of the first recruits was Geoff Crompton, who arrived from Rootes in 1960, bringing his wife Pauline with him as one of Rover's first colour and trim designers, and continuing the role she previously played at Rootes.

Maureen Hill arrived as Bache's secretary in 1961, having joined Rover in 1958 on completion of her secretarial training. "The design studio was not large, but had the showroom and the modelling shop adjacent" she recalls. "The showroom had a full-size turntable and a lift that brought the vehicles up from the engineering workshop and also could take them up onto the flat roof for photographic sessions."

At this point there were around a dozen modellers who worked in different materials, making the wooden armatures for clay models, led by chief model maker Ken Barton, who had replaced Frank Underwood. They also constructed interior bucks to show the proposals for instruments, seating and trim, working in conjunction with the trim shop and other engineering departments. In addition to the Cromptons and Tony Poole, the other designers were Francois Talou, Jim Hirons and Norman Morris.

The Cromptons later encouraged William Towns to join them from Rootes, which he duly did, arriving at Rover in 1963 and staying for three years. Talking in 1983, Towns said "The only car at Rover I could claim authorship for was the second Le Mans gas turbine racer. On the production cars I did a lot of detail stuff: knobs and instrument panels and switchgear."

Geoff Purkis joined Rover in April 1964, aged 23. He had trained at Birmingham College of Art and subsequently worked at Lucas as a designer. "I knew Rover from my college days where Vic Roberts was a day release student from the studio," recalls Purkis. "He was a very creative chap, and later did a nice series of designs for a modular Range Rover design with different back ends. I was interviewed by Bache, but there was no vacancy at that point. Then a year later I joined Lucas and learned how to draw realistic lamps in airbrush. I went to see Bache again on Lucas business whereupon I got this letter asking if I was interested to come and work for him."

"There was quite a bit of 'old school tie' around in the Wilks days. Like David, I'd been to an independent school and whether he thought that would make me fit in I don't know. Or did he just like my airbrush work? – who knows!"

Another new addition in 1964 was Len Smith, who applied for a job as a 'model maker' after having seen an advert in the *Birmingham Post*. Smith's background was in patternmaking and tooling so was quickly hired by Ken Barton. "When I first joined there was a beautiful full-size wooden model of a P5 in the studio. I was stunned, I had never seen anything so elaborate in my life and I had no idea that people were employed to make such wonderful creations. It was a whole new world to me." He confirms that they used Milton Bradley clay from America, the English clays such as Wilkins Campbell being too waxy and soft for good modelling work. Smith went on to become Barton's assistant and stayed with Rover throughout his modelling career, becoming modelling manager at the Canley and Gaydon studios into the 1990s.

Rover in the mid-1960s was one of the best-run British car companies, with a solid management team that was dedicated to advanced engineering and increasing passenger safety in cars. With Bache at the helm in design, it was one of the first companies to employ proper ergonomic principles, also conducting research into seating comfort in interiors. Bache embraced the use of new materials for luxury cars, and made a point of informing himself of the latest supplier technologies that he might encourage his team to employ in their work. This was in marked contrast to Jaguar, where

Bache and his 34-strong design team, late 1960s. (Courtesy Maureen Hill)

the traditional veneered instrument panel and crude armatures persisted well into the 1970s.

Bache nurtured a thoroughly modern approach to design that reflected contemporary trends in architecture, product design and graphics, meaning his team produced high quality work that did not slavishly follow the fashions of the Ford or Vauxhall studios at the time, and was superior to that of the smaller design studios such as Triumph. In outlook it was closer to the ethos at Volvo or Saab than Humber or Jaguar. Here was an independent British company that could embrace toying with advanced gas turbine research and developing military tanks under its Alvis subsidiary, while introducing a smooth, powerful American V8 engine to spice up the sedate image of its P5 and P6 saloons.

"David Bache was very interested in industrial design as such," said Towns. "At Rootes designers didn't do much more than draw the patterns which were heat printed in the vinyl, but at Rover you found that Styling would actually play a great part in the construction of a seat."

Two further apprentices joined in 1968. Chris Wade and Kevin Spindler had both started as Rover apprentices in 1964 and on completion of their studies entered the studio. "Amazingly there were few takers from the apprentices to go into Styling and Kevin Spindler and I went in together," recalls Wade.

"Jobs were meted out," he continues. "One thing that was brilliant about David, he was a great salesman. He was always pushing the boundaries. One was allowed to do sketches of drag racers and dream vehicles. At first I'd hide them as he came past, but later I learned he actually liked that stuff and would comment on them, saying 'keep that feature, this is interesting, don't let that idea go' etc." Purely off his own initiative, Wade did a series of Range Rover buggy sketches and got the model shop to make a second GRP shell of the scale model made for the Louvre exhibition in 1971 with an open pick-up rear.

"I think having Land Rover within the range helped Rover design stay focused on industrial design rather than pure chrome trim and tailfins," he concludes.

By the late 1960s, Rover's styling department had grown to around 30 staff. By now, Geoff Crompton had departed, Hirons had gone to Ford in Germany, and Towns had moved on to Aston Martin. In their place new arrivals had joined, including jewellery designer Ian Beech and Graham Lewis for colour and trim.

Range Rover development

The Road-Rover idea never went away, but engineering resources were lacking and needed to be focused on other projects. Firstly, there was the Series II Land Rover for 1958, plus the big P5 saloon. After that date the main priority was P6, but there was also the P5 Coupé which debuted in autumn 1962. Thus, it was only after 1963 that attention could be refocused onto new projects that Spen King was keen to pursue.

Rover styling studio. By the mid-1960s, the styling studio on the first floor had expanded into Gordon Bashford's old engineering design area, with each designer having their own penned-off area. Sketches in marker pen, pastel and gouache on Canson paper were the order of the day. Tony Poole stands on left. (Courtesy Len Smith)

The motivation for the Range Rover project came from two sides. The first came in 1965 from Graham Bannock, in charge of Rover's market research department. One of his first tasks was to visit the US, where Rover was struggling to sell both its saloons and the Land Rover. He noted the success of the International Harvester Scout 80 from 1961 and the Jeep Wagoneer, introduced in November 1962, both of which were more civilised versions of the basic Jeep 4x4. More significantly, Ford entered this compact leisure 4x4 market with the launch of the Bronco Wagon in August 1965. This offered a smooth in-line six-cylinder engine, with the option of a 289cu in V8 to follow in March 1966.

Bannock thought that Rover could produce such a recreational and leisure vehicle, and produced an outline description of his thoughts in a memo. Coincidentally, his thoughts followed the kind of dual-purpose vehicle that Rover engineers were also considering at this point. After witnessing the Bronco launch in August 1965 he returned to Solihull, where he met with Spen King, in charge of New Vehicle Projects. King and Gordon Bashford discussed their ideas

Land Rover Design – 70 years of success

Ford Bronco (above), International Harvester Scout (right), and Jeep CJ-5 Wagoneer (below). The Wagoneer was more luxury-focussed, while the Scout was very basic, comprising updated 1960s styling on a Jeep-type chassis. The Bronco found a sweet spot between them, offering two-door hard top, pick-up and novel open-door 'Roadster' versions.

David Bache era and Range Rover

with him, and the pair then set about putting together the rationale for the project that autumn. Technically, the idea was to combine full-time 4x4 and quite soft, very long travel suspension – quite the opposite of the Land Rover. The recent acquisition of the rights to build the Buick aluminium 3.5-litre V8 engine would provide a lightweight power unit with a creamy power delivery and loads of low-down torque – perfect for their new concept.

The second motivation was political. After the election of Harold Wilson's Labour government in April 1966, the purchase tax on cars was increased, leading to a downturn in sales. Not only that, but defence spending was cut, and orders from the MoD for Land Rovers were reduced. Rover was being squeezed on both fronts, and the management needed to come up with a strategy to combat this new reality. Suddenly, Spen King's NVP strategy gained a new impetus.

They soon had a basic layout scheme on paper, and 20 engineers were seconded to work on this exciting new project, initially termed the 'Alternative Station Wagon.' Geof Miller was one of the engineers seconded to NVP, joining the project in July 1966. Nick Wilks, then a technical assistant in Engineering, was also seconded to help, together with chassis engineer Phil Banks, and body engineer Phil Jackson.

In autumn 1966, Miller arranged a 4x4 comparison test at Eastnor Castle testing grounds. Examples of the Ford Bronco, Jeep Wagoneer, and International Harvester Scout had been purchased by the Rover NA subsidiary and sent over for evaluation. All were two-door bodies, and so the assumption was that this was what the US market demanded. It would also allow simpler CKD assembly overseas, following current Land Rover experience. The Eastor Castle session was followed by trials of early permanent 4x4 drivetrain mules in the snowy lanes around Edgehill in early 1967.

The first prototype 100/1 was produced in the Mock-Up Shop in January 1967. Although Spen King is widely credited with styling this first model, more recent research reveals he had some unofficial help. According to King "David refused to have anything to do with

Newly-completed prototype 100/1, July 1967. The body style shown here was sketched out by Geoff Crompton as a temporary measure in order to clothe the running gear, before Styling devised the definitive design. However, management liked this proposal so much, it asked for it to remain, with only the lightest of changes.

Land Rover Design – 70 years of success

Prototype 100/1 registered SYE 157F was painted mid-grey, seen here with a later style of mesh grille. The bumper and wipers were both taken from the Ford Transit. Prototype 100/2 was LHD, painted mid-blue and registered ULH 696F in April 1968.

Rear view, same day in spring 1968, showing the messy spilt tailgate arrangement with double locks on each tailgate door. Neither of the prototypes have survived.

David Bache era and Range Rover

Interior of 100/1 featured a double passenger seat in the front.

Styling input officially began in June 1967, with scale model proposals based around the 100in package developed by NVP. These followed more car-like and less boxy themes favoured by Bache but did not proceed beyond quarter scale clay stage. King pointed out difficulties with CKD assembly with these more elaborate themes, so a second exercise was instigated to simply tidy up Crompton's 100/1 body style.

Like many examples of great car design, it is fascinating to note how this classic design was done very swiftly over an intense period of weeks, starting in July, and completed by September 1967. Comparing it to the Crompton style, Bache and his team modified the exterior styling with a series of carefully considered changes, introducing a very fresh, architectural theme that was to prove a timeless design.

Firstly, they introduced a recessed bodyside feature, rather than the double 'clinker boat' feature lines of the prototype. Added within this recess were all the functional elements of the design, including front and rear lamps, turn signals, grille, fuel filler and vertical door handle. Attention then turned to the front, where a broad black grille with slotted vertical lozenges was incorporated. The greenhouse and bonnet were left pretty much untouched, except for a pair of raised castellations added on the front tips of the bonnet to provide some corner guide points when driving off-road, and as a mounting area for the wing mirrors.

The rear end required the most revision, with the awkward exposed weld flanges replaced with proper steel corner pressings.

it for a long time. Initially we designed it ourselves. We borrowed a stylist – without David's knowledge – called Geoff Crompton, who helped us on the drawing board. Otherwise the Range Rover came straight out of our outfit [NVP]. It didn't go into Styling except for tarting it up at the end."[1]

This is crucial. Crompton's input meant the 'Alternative Station Wagon' possessed some fine basic proportions and an aesthetic balance that the previous Road-Rover had sorely lacked. The low belt line and generous glass area echoed contemporary styling fashion, and the construction of the greenhouse pillars obeys many classical rules of car design, with the A-pillar pointing directly to the centre of the front axle, and the other pillars converging to a single point above the car.

Into 1967, the project became known as the '100-inch Station Wagon' in internal memos. By July, prototype 100/1 was ready and Technical Director Peter Wilks was delighted with progress so far. A formal product proposal was drafted and presented to the Rover board that same month. This was accepted, and a fully detailed plan was approved in January 1968, with a production aim of December 1969 – a very fast programme indeed.

An alternative scale clay model for the 100in Station Wagon was produced by Styling. Viewed today, it looks over-styled and not as rational as the final Range Rover.

1. According to Maureen Hill, one of the reasons for Bache's reluctance to get started on the Range Rover was that he was busying himself on a personal project for a P6 Coupé. The car, subsequently known as 'Gladys,' was being built at the coachbuilders, Radford, and was an unofficial proposal for an Alvis. The prototype cost £22,000 but was bought from the company by Bache, and subsequently used by his wife Doreen for many years.

Land Rover Design – 70 years of success

Rear view of same double-sided model, with overtones of the new Renault 16 hatchback. Bache wanted a more car-like vehicle, with a less utilitarian look, but in retrospect it would have dated quickly.

Full-size model sitting on the turntable in the presentation room on the first floor. The white drape above was used to diffuse the harsh strip lights. The vertical door handle does not yet feature. This clay model was recreated in 2016 as part of the new visitor experience exhibition at Lode Lane. (Courtesy Maureen Hill)

It is believed just this one full-size clay model was produced for the Range Rover development in July 1967, with alternative themes on each side. The vertical grille bars and chunky 'capped' turn signal lamps have appeared, together with deeply recessed 7in headlamps. Note the 'Road-Rover' script and that the far side of the bonnet has no castellations. (Courtesy Maureen Hill)

The upper tailgate glass stayed as a flat sheet of glass with a simple frame supported on gas struts, while the lower tailgate was cleaned up with a deep undercut to provide a surface for the single release handle and the licence plate lamps. Rear lamps were redesigned to match the front, as a chunky lighting block that wrapped around the corner of the car, sitting within the recessed bodyside feature.

It is intriguing to understand how other 'iconic' features of the Range Rover came about through functional constraints. The low belt line and slim pillars were typical of King's obsession with good visibility. Like the earlier Road-Rover, the sliding rear side glass demanded that the belt line and cant rail ran dead parallel. The 'floating roof' panel was bolted on with an exposed roof gutter, not unlike the BMC Mini.

In truth, there wasn't much spare time for the 100in Station Wagon project that summer. The main focus of Bache's time was focused on the big P8 saloon car project, which was planned to replace the P5 and was the major programme for the Rover Company. Not only that, but there was the exciting mid-engined P6BS sports car, codenamed P9. This was the other pet project of the NVP team led by King and Bashford. A single prototype had been built concurrently with 100/1 in late 1966 and was running in early 1967 using a simple body design comprising flat planes, with the intention of having styling input once the project was approved with Rover's subsidiary company Alvis. Once again, Geoff Crompton was seconded to provide a semblance of styling input to Spen King's layout, producing another unfussy bodyshell design for the prototype.

By September 1967, 100/1 was undergoing test trials, with excellent results. The long-travel coil spring suspension provided exceptional wheel articulation and the strong 3.5-litre V8 engine with permanent 4x4 provided plenty of power, even in its de-tuned 130bhp form. The solid axles used radius rods on the front, as on the Bronco, but the layout gave terrible steering kickback at first, requiring a swift redesign to cure it.

David Bache era and Range Rover

Alternative front end detailing on a pre-production Velar in the studio. (Courtesy Maureen Hill)

Land Rover Design – 70 years of success

Tony Poole pours over Range Rover sketches, summer 1967. Graham Lewis (with beard) is seen in the background. (*Style Auto* magazine)

The interior design was a totally fresh start, with Tony Poole taking the lead on its development from early 1968. Occupant safety was an important consideration, and the double passenger seat as tried on 100/1 was abandoned in favour of two high-strength safety seats, with a built-in lap and diagonal harness. The rear seat bench was designed to fold forwards to provide a huge trunk capacity: 1.5 metres long and over 1 metre high.

Bache kneels to look at sketches (above), and shares a joke with modeller Harry Crudgington (left). Rover modellers were issued with blue smocks (Courtesy Maureen Hill)

David Bache era and Range Rover

Other safety features included a collapsible steering column, burst proof door locks, collapsible window winder handles and a high-resistance moulded GRP headlining, affording greater occupant protection in the event of a collision.

The plan was to offer a basic interior and second De Luxe version with rear-facing trunk seat for children – not unlike the Volvo Amazon estate. Prototype 100/7 was built with this latter feature, but, as the production date loomed nearer, the priority switched to getting the basic interior finalised in something of a hurry, so the De Luxe idea was abandoned.

Styling continued to develop and finalise all details throughout 1968-69, with the small team also busying itself on the face-lift and redesign of the P6 for 1970, as well as continuing to develop the big P8. Tony Poole split his time between the Range Rover and Land Rover Series IIA updates, where new legal requirements for certain export markets required the headlamps to be moved outboard onto the front fenders in spring 1968. To alleviate fears of vulnerability, they were housed in slightly recessed pockets with the turn signals and sidelights now mounted vertically to complete the neat composition with the enlarged rectangular grille mesh. UK lighting regulations soon required this update

A Tony Poole background with a pasted-on photo of the clay model. Poole used this photo-montage method for much of his detail design work. (Courtesy Maureen Hill)

Production seat arrangement with integrated seatbelt and styled ABS endcaps on the seat plinth. The front seats had to withstand a pull of 7000lb ft in a crash situation, something that had never been attempted before and was not easy to engineer.

Styling mock-up of the interior. The front seatbelt buckle is mounted on the seat, but the reel remains on the B-pillar, hindering rear access. Note the console tray proposal. (*Style Auto* magazine)

Land Rover Design – 70 years of success

too, which was introduced across all Series IIA Land Rovers from February 1969.

Company politics were also playing a part. In 1966 the Rover Board finally decided they needed to merge with another car company to achieve the critical mass required to withstand the increasing competition and vast levels of investment that were now needed for new model development. After protracted talks they were bought by Leyland Motor Company. Once the new Leyland bosses Lord Stokes and John Barber were allowed to see the '100in Station Wagon' in summer 1967 they immediately sanctioned it for production, swiftly realising its unique potential.

At this point, the P8 saloon was proceeding steadily, as was the P9 sports car, but inter-company politics would see both projects re-examined once the next round of mergers was concluded. In January 1968 it was announced that Leyland Motors would merge with British Motor Holdings to form the vast new British Leyland company and – somewhat surprisingly – the P8 saloon was given the full go-ahead for production as a new 'Mercedes-beater.'

Nevertheless, work was still needed on the Range Rover and P6 saloon. Young development engineers Roger Crathorne and Alan Wood joined the NVP team early in 1967, and a second prototype 100/2 was built in spring 1968, this time LHD. At this point Spen King was removed by Lord Stokes to head up Triumph Engineering, and in his place Gordon Bashford was promoted to head up NVP, although it was mainly left to Geof Miller to complete work on the 100in Station Wagon.

"We drew components properly and then made them rather than just making components on the hoof on the workshop floor," says Crathorne. "Tom Barton was an old-fashioned railway engineer, convinced that leaf springs were best for off-roaders. He took a lot of convincing otherwise by Spen. He thought it too softly sprung, would be prone to chassis cracks (it wasn't)."

Barton's philosophy for the Land Rover was simple: keep the vehicle's ride harsh and uncomfortable as a means to limit the strains put upon it by the thresholds of the occupants, and the concept behind the '100in Station Wagon' was a direct challenge to this view. Miller confers: "Barton was not at all keen on the project. Coil springs: ridiculous. Disc brakes: ridiculous. Barton wanted a hose-out interior too. I felt it should be more upmarket but lost that argument. The untrimmed boot was too sparse."

But what to call it? Miller confirms that marketing searched for several names for the vehicle including 'Panther' and 'Leopard.' Other names being seriously explored during 1968 included Land Rover 'Ranger' and Land Rover 'Viking,' although the latter name was owned by GM and permission would have to be sought to use it. Finally, Tony Poole suggested 'Range Rover,' and the satisfying alliteration and adventurous associations of the name compared to 'Road Rover' were easy to appreciate.

So, Range Rover it was. The Rover Technical Directors' Report of 25 September 1969 stated the following: "It has now been decided

Roger Crathorne

Roger Crathorne has been deeply involved with Land Rovers for over 50 years, and these days acts as a celebrity ambassador for JLR, with a huge fund of knowledge of the brand heritage. He was born in Solihull on 8 April 1947, in the same Easter week that Maurice Wilks drew the image in the sand. Solihull in the 1940s was an affluent village rather than a town, and Crathorne was familiar with the company just down the road. His father worked at Rover in Purchasing, so it was no surprise that Crathorne started an apprenticeship there from September 1963, aged 16. He initially worked in the Engineering office on the 110 Forward Control, working under Tom Barton.

"At Land Rover in those days, you were not only an engineer, but were also a designer, a developer and a prototype builder. You did everything. More so than at Rover Cars where you didn't tinker and bend metal like you did at Land Rover. Rover cars were designed and conceived and packaging was important: Land Rovers were engineered much more on the hoof."

In late 1966, Bernard Poole recommended him from the apprentices to be one of the junior engineers for the 100in Station Wagon project that was starting in NVP. Chief project engineer Geof Miller selected him to join the team. Crathorne did a lot of the off-road development testing on the project and evaluated competitors including the Jeep Wagoneer, Toyota FJ55, Ford Bronco, IH Scout and the big IH Travelall. Crathorne was one of the demonstration drivers at the Range Rover launch at the Meudon Hotel, Cornwall in June 1970 and features in many of the launch photos that were publicised at the time.

Crathorne continued to work for Miller during 1970 on developing the Range Rover after the launch, and by 1972 they had built an early four-door prototype – but were told to stop work by management and the vehicle was rebodied.

After being involved with the build of the hybrid Stage 1 V8 Land Rover prototypes in 1976, he became manager of the Land Rover demonstration team from 1978. With a team of five people under him providing off-road driving advice at the Solihull jungle track and Eastnor Castle, he became something of a Land Rover celebrity as the PR team started to use him increasingly for press events. During the course of a year, over 10,000 people would experience the jungle track.

With the establishment of the Land Rover Experience in 1990 Crathorne was in regular demand with the media, and paying customers for off-road tuition and guidance, plus his seal of approval for off-road capability on new models, has become a valued ritual. Crathorne retired after 50 years' service in 2013.

Roger Crathorne seated in HUE 166.

David Bache era and Range Rover

that the 100in Station Wagon is to be known as 'Land Rover – Range-rover' (sic), the words 'Land Rover' to appear at the front and the words 'Range Rover' on the sides." By the time of the launch, the final badging showed 'Range Rover' on the front fenders and bonnet and a 'Range Rover – by Land-Rover' label on the rear tailgate.

The resulting final design by Bache's team was a masterful retouching of the original prototype. The front end was bold, the single round headlamps flanked by oversize chunky sidelights and turn signals, giving a confident, non-aggressive expression to the face. The door handle appearing as a vertical black rectangle set within the bodyside recess epitomised the confident use of graphics on the car, and was a brilliant combination of contemporary design and functional ergonomics. As the first contact point for the customer it provided a rugged introductory 'handshake' to the vehicle, being large enough to operate with thick gloves and having a solid, well-engineered action.[2]

Brightwork was kept to a minimum, with matt black being the

---

[2]. By contrast, the later four-door Range Rover had one of worst-engineered door handles of its day, inherited from the Morris Marina.

Publicity launch photo of YVB 153H, the Tuscan Blue pre-production vehicle that featured in the original sales brochure.

Land Rover Design – 70 years of success

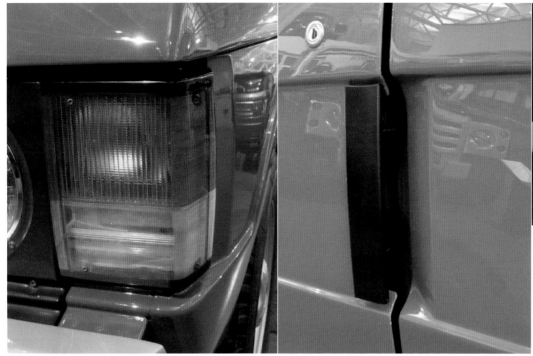

Characteristic details of the Range Rover include the chunky turn signal lamps, vertical door handle flap, and the Rostyle wheels. (Author's collection)

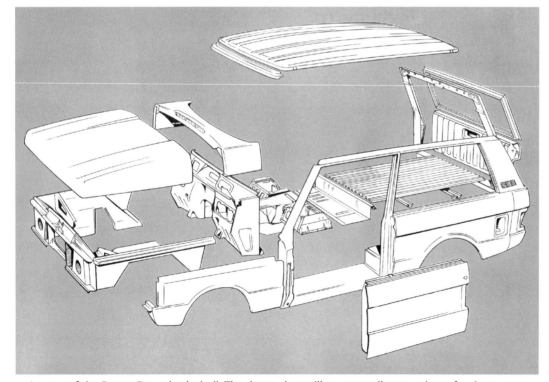

Layout of the Range Rover bodyshell. The doors, door pillars, rear tailgate and rear fender corner panels were made of steel; other skin panels were aluminium. The bonnet was originally meant to be aluminium but addition of the castellations meant it could only be pressed in steel. Spen King stated he didn't approve of the change! (*Style Auto* magazine)

dominant exterior finish. This followed the latest fashion, influenced by rallying and motorsport, and would be adopted for the P6 face-lift being carried out at the same time for introduction in 1970. Only the wipers and washer nozzles had a bright stainless steel finish (although these were later changed to black), with the bumpers being given a simple silver paint finish.

The body construction was innovative, too, using a steel base unit, as on the P6 saloon, with Birmabright aluminium outer panels attached by rivets. Steel was used for the bonnet, door pillars, tailgate and rear fender corner panels.

It just needed one final touch. Rubery Owen had developed a new design of wheel that avoided the need for fussy hubcaps. This styled wheel used a new deep-draw manufacturing process for steel wheels, and was marketed from 1967 as the 'Rostyle' wheel. Rover was one of the first adopters of the Rostyle wheel for the P5B, as was Ford for the Cortina Mk II 1600E, both using an expensive chrome finish, and Bache decided it was exactly right for the Range Rover, albeit with a simple silver paint finish. The use of this latest sports-style wheel with knobbly 205/16in M+S tyres was a conclusive detail that added to the 'Tonka Toy' appeal, and set it apart from any other 4x4 vehicle.

David Bache era and Range Rover

Front cabin of YVB 160H with seats removed. One can detect the hurried execution of the design, with the instrument panel being an assembly of plastic panels and the footwell area being crudely trimmed with rubber mats. The tunnel console was deleted, leaving just the ashtray isolated on the rubber-covered tunnel.

YVB 160H in the photography studio for the launch brochure shots. Evident here is the sparse trunk trimming and door linings with their double-ended door grabs and twin release handles. The box-pleated trim style was achieved using latest vacuum-forming processes on foams, but proved not so durable in service, and is nearly impossible to recreate today for a restoration.

Land Rover Design – 70 years of success

Trunk layout of the same pre-production vehicle showing the sparse trimming with simple ribbed mat on the floor only. Early models had a fold-down hinged licence plate.

This combination of contemporary functional design with a spark of sportiness was carried through to the interior. The IP was designed as a non-handed ABS plastic moulding, with the instruments sitting in a small add-on binnacle that was matched to an infill panel on the passenger side, creating a storage recess. Optional supplementary instruments were sited in the centre of the car as individual dials, as on a sports car. Controls were large and chunky, typified by the manly gearlever, but the use of moulded rubber for the two gearlever bellows and floor mats and a large-scale grain for the plastics provided a robust authenticity that was highly appealing.

On opening the door to the cabin the seats were the first thing one noticed. Mounted high up on a plinth, they were trimmed in a vacuum-formed PVC material with wide box pleating: a style later introduced on the P6 saloon. The sides of the plinth were capped with a neat ABS moulding that was typical of late-1960s product design: a trapezoid form with recessed detailing and radii that perfectly matched the seat above. The integrated seatbelts allowed unimpeded access to the rear seats when tipped forward, together with a correct ergonomic relationship to the occupant, no matter where the seat was positioned.

Only one interior colour was available – Palomino Beige – which contrasted nicely with the six exterior colours offered, and provided a more upmarket and modern ambience than the plain grey or black PVC on the Land Rover.

Launch of the Range Rover

After the formation of British Leyland, Chairman Donald Stokes gave a commitment to launch two new models every year. For 1970 he had the Triumph Stag, but lacked another launch, hence the Range Rover – by now due for spring 1971 – being brought forward to June 1970.

Just seven initial prototypes had been made, the first two with the Crompton body. Then 25 VELAR prototypes were built for development. Testing went well – not only in off-road, but also in customer clinics.

David Bache era and Range Rover

Orange fun buggy sketch by Chris Wade. Wade did a number of Range Rover sketches that were exhibited at Lode Lane in 2016. (Courtesy *Style Auto* magazine)

Rather than the original plan for a Morocco press launch, the Range Rover was presented to the press in Cornwall on 17 June 1970. Stokes had wanted it to be even earlier – April 1970 – but the required number of pre-production vehicles were simply not available in time.

Press reaction was overwhelmingly positive. The Range Rover captured the imagination of every journalist who drove it, with the result that demand was immediate and sustained – customer waiting lists were drawn up as soon as the Range Rover appeared.

The situation was simple: the Range Rover was launched at a price of £1998, and, at the time, there was no opposition that could offer the breadth of ability that it possessed. Not only was it a highly accomplished off-roader, but it was also a commodious estate car, and (as Rover would soon find out) something of a status symbol. Customers loved the command driving position, and although farmers and van drivers might have been familiar with this, to the buyers of prestige estate cars such as the Volvo 145 or Triumph 2500, it was a completely new experience.

Very soon, Rover realised that people were buying its new flagship for many reasons other than its off-road capability. The 'four-cars-in-one' idea of farmers using it to carry sheep and then hosing it out to go to the theatre in the evening slightly missed the true customer, which proved more likely to be the landed gentry and style-conscious middle classes living in London, leading to the soubriquet 'Chelsea tractor' being coined by the late 1970s.

The Middle East was another key market; Arab buyers loved the associations with the British royal family, game shooting and polo matches. They also loved the superior seating position, even if they did not like the idea of just two doors for being chauffeured around. The police and emergency services also quickly realised the potential of the Range Rover for high speed motorway patrols, where the 95mph top speed, raised driving position and ability to shift a 40-tonne truck off the carriageway were key assets that no other vehicle could provide.

The Range Rover went on to be a huge success throughout the 1970s, in spite of British Leyland's dire management. The company's lack of development on the Range Rover was shocking, but in reality – and like other models in the BL range, the Mini or the MGB for example – its underlying integrity would allow the company this neglect. It had to be this way, because BL were fighting a huge battle in the middle of the market with the volume cars, where the majority of sales were to be found. The Range Rover would have to fend for itself. Fortunately, customers continued to buy it, and did so because it was such a unique vehicle, even if build quality was patchy and delays in production meant long waiting lists.

Bache and changing responsibilities in design

Work on the Rover P8 saloon commanded the majority of the Rover studio resources throughout 1968 and 1969, as did the P9 sports car but inter-company politics would see both projects canned at the start of the new decade. Development work on both P8 and P9 Sports continued up to summer 1970, with £3M of body tooling for the P8 having been spent with Pressed Steel Fisher. At this late stage Sir William Lyons at Jaguar demanded that both projects be axed, seeing them as too threatening to his own XJ6 and E-type models. While work on the P9 ceased immediately, work on the P8 continued until March 1971 when it was terminated.

Land Rover Design – 70 years of success

It was a sad day for Solihull and Bache's studio must have been gutted to lose two major projects so soon after receiving awards for their work on the Range Rover, not to mention the excellent reception for the face-lifted P6 introduced in September that year. A sense of anticlimax prevailed throughout the studio and, for the first time, Bache must have wondered whether his tenure at Rover was totally secure.

He need not have worried for long. Although Bache lost out with the axing of the P8 and P9, his personal career path was looking bright as 1970 moved into 1971. He was now promoted to Design Director of British Leyland's Specialist Division, incorporating sports cars, luxury saloons – and Land Rover. In this new role he would eventually manage the Triumph studio at Canley, with design work for two new ranges of saloons being planned.

The first project was to replace the Rover P6 and Triumph 2000/2500 ranges with a single model, eventually to become the Rover SD1. Design work on a P6 replacement – the P10 – had begun in summer 1970 and quarter scale models were approved for further work in February 1971 – just prior to the P8 being axed. By July the full-size clay for this SD1 programme was under way and was ready for approval by November – although it took a further five years to get it into production, such was the muddled bureaucracy at BL at the time.

The second project that occupied a lot of resource in the period 1972-75 was for a smaller SD2 model to replace the Triumph Dolomite/1500 range and compete with the BMW 3-series, Alfa Romeo Alfetta and Audi 80 in the compact prestige market. Finally, there was the replacement for the BL sports cars, culminating in the Triumph TR7, launched in 1975. For all of these programmes there was a great deal of inter-studio competition with Harris Mann's Austin-Morris team at Longbridge, not to mention external proposals from Michelotti and Pininfarina in Italy.[3]

---

3. *One of the designers in the Triumph studio was William Towns, formerly employed by Bache at Rover. He had left to join Aston Martin, and subsequently set up as a freelance designer, but had struggled for a time. He was then re-employed as a consultant to Triumph working on proposals for the joint Rover/Triumph programmes, but his model lost out to Bache's model for the SD1, and by 1973 he had left the company to resume his freelance activities.*

Tony Poole

Tony Poole started out as a fitter in Rover's gas turbine department, having been involved with jet turbines since his days in the RAF in the latter days of WW2. While in the RAF he had a near-fatal accident when a propeller sliced into his face, giving him a permanent deep scar on his right cheek.

Poole was recruited by David Bache as his first styling assistant in spring 1956. He had sketched suggestions for badges and other details on the Rover T3 gas turbine car that caught Bache's eye, and on the strength of this found himself a permanent job in the styling studio. Although not a professionally trained designer, Poole's skill as an illustrator using an airbrush was highly acknowledged by other colleagues.

"I liked Tony. He was quite eccentric," recalls designer Alan Sheppard. "Some of the art work Tony had done was astounding, the most amazing airbrush renderings I've ever seen. Variations on tail lights, door release handles – very detailed. So real you could take them out of the paper and understand how they worked."

Poole's first styling job under David Bache was to design the revised roof profile for the P5 Coupé, which was then modelled as a hardwood body former for surface release to Pressed Steel.

In the late 1960s he split his time between the Range Rover and Land Rover Series IIA updates, where new legal requirements for certain export markets required the headlamps to be moved outboard onto the front fenders in spring 1968. Poole also sketched out proposals for a new Land Rover One-Double-One project, a proposal for a restyled Land Rover on a 111in chassis. Although the idea was presented to management, it was not pursued. Poole is also credited for coining the name 'Range Rover' itself, with the title formally adopted at a meeting on 18 December 1968, when the project team met representatives of the Rover Sales Department to discuss a number of issues.

Tony Poole was an accomplished cartoonist, and soon found his talents in great demand around the company for humorous retirement gifts and Christmas cards. "In his latter days he was frustrated, some of his liberties had been taken away," continues Sheppard. "You might think him quite sad, but he always managed to present a jolly outlook. Just before he retired he went through a period where he locked himself away in his office, we thought he was going to leave us some big project. We could hear Magic Markers, paper being cut, the occasional clink of a gin glass. More cutting noises. This went on over several days. His office wall didn't quite reach to the ceiling and in front of it was a long plan table down the centre of the studio. As I sat there, suddenly a huge shadow covered my work station. There was a whoosh noise and as I looked up an enormous Foamcore biplane flew over and landed on this plan table. A perfect landing. The pilot was an effigy of himself, complete with a Biggles scarf ..."

Tony Poole with his characteristic pipe that was never far from his side. Seen here in conversation with Len Smith (centre) and David Bache. (Courtesy Maureen Hill)

Series III and SD5

With Bache having wider design responsibilities, design work for Land Rover took a back seat for much of this period in the 1970s. Tony Poole took on responsibility for any Land Rover modifications and was behind the styling upgrades that lead to the Series III, which debuted in 1971.

Apart from the plastic grille, the biggest change was to the interior, where new legislation required the instruments moved to a position in front of the driver, housed in a small binnacle. A padded top roll was added beneath the windscreen and a full-width shelf formed from simple ABS vacuum formings provided a more car-like interior, complete with a plastic shroud for the steering column and

Series IIA Land Rover Station Wagons with outboard headlamps, introduced across all Series IIA Land Rovers from February 1969. When first launched, many people felt it gave the vehicle a slightly wide-eyed 'ovine' expression compared to the original 'Landie.' Familiarity over the years has mellowed that impression.

a new column stalk for the horn, headlamp flasher and direction indicators. A proper fresh air heater was incorporated for the first time, with the inlet let into the nearside front fender.

Tony Poole was promoted to 'Executive Styling Engineer' for Land Rover in 1973, the first time the sub-brand had any dedicated styling design resource. He subsequently contacted Chris Wade – who had left Rover some months previously – to ask him back to join him as the second designer on the team. Wade turned down the offer and the position went to Charles Coldham instead.

Coldham was an unusual addition to the team, a highly cultivated, slightly nervous character, but capable of producing some beautiful artwork. Some years later, as a more senior member of the design staff, he was allowed a BL company car and – being single – promptly blew a considerable chunk of his monthly salary leasing a Jaguar XJ-S, much to the annoyance of the other Rover executives with their SD1s.

Once the Series III was out of the way, Land Rover engineers turned their attention to a complete update for the Land Rover, codenamed SD5. Designer Steven Ferrada joined Poole and Coldham to come up with a neat design using a modular construction around a set of base units that would allow simple variations of wheelbase, hard top and pick-up bodies. This 'modular dray' concept was even envisioned

Land Rover Design – 70 years of success

Series IIA interior featured a De Luxe seat option (left) with fluted PVC panels and a modicum of side support.

Series III featured flat door hinges and an intake vent on the nearside fender for the new heater. Inside was a new IP with the instruments in front of the driver, plus a new dished steering wheel with a plastic cowl.

David Bache era and Range Rover

Series III range was introduced in 1971 with a new ABS plastic grille that aped that of earlier Land Rovers, such as the Royal parade vehicle from 1958.

Scale model of SD5, early 1975. The design was notable for the bold graphic break-up and use of a textured scuttle panel. The panels at the top of the wheelarches were flexible rubber inserts that could accommodate different wheel and tyre combinations. (Courtesy Maureen Hill)

as a sporty version, too. GRP components were considered for the Hard Top and other exterior parts, while the seats were designed as modular blocks that could have a variety of different insert pads installed.

With such an open format of vehicle, the styling bucks were done in wood and clay, to produce a simple mock-up that could be pushed around and viewed outside. In overall style it was not unlike the Fiat Campagnola and Mercedes G-Wagen that were being developed around the same time, and the SD5 could have provided an earlier third product line than the Discovery that came some 15 years later.

In the end, the style was judged to be too close to the Range Rover, plus the project coincided with another financial crisis in BL: a new Land Rover was a long way down the priority list for the cash-strapped company. This was a pity, for the SD5 could have pre-empted the trend towards lifestyle 4x4s in the late 1970s capitalised upon by the Japanese – particularly Suzuki and Toyota.

For the Range Rover there were minor design and equipment changes. In October 1973, Bri Nylon cloth seats became an option, and the C-pillar was clad in black vinyl in a slightly desperate attempt

Continues on page 62

Land Rover Design – 70 years of success

Presentation of SD5 in the styling studio, with sketches by Steve Ferrada. The modular concept allowed a number of different body variations to be built around a basic cabin. (Courtesy Maureen Hill)

Same day, early 1975. A rudimentary package mock-up made of plywood. (Courtesy Maureen Hill)

SD5 was made as a clay and wood mock-up for viewing outside, finished in green. The axles appear to use a Range Rover-type location, not leaf springs. Note the lack of interior. The Futura Black typeface used on fenders and bonnet was a fashionable 1970s choice. (Courtesy Maureen Hill)

Same model, photographed with the new East Works construction in the background. The date is 9 October 1975. (Courtesy Maureen Hill)

Land Rover Design – 70 years of success

The SD5 and Fiat Campagnola that appeared in 1974 both followed a similar product design aesthetic.

The Renault Rodeo of 1970 was another light leisure off-roader that followed a modular product design aesthetic like the SD5, with vertical graphic elements finished in a contrast colour. Pictured here is the Series 2 Rodeo from 1981.

to hide the poor quality of the underlying pressing. By 1979 there were minor cosmetic changes: bumpers were now painted black, not silver, rectangular door mirrors were fitted in place of round wing-mounted ones, and tail lamps were revised to incorporate mandatory rear fog lamps. On the interior, the steering wheel was now a four-spoke design, rather than the three-spoke version of earlier models, and factory-fitted air-conditioning could also be specified for the first time.

Land Rover splits from Rover

Once the SD5 project was canned, little would change product-wise over the next few years. For Land Rover, the 1970s were something of a lost decade, and any investment carried out by British Leyland was diverted towards the mainstream car ranges. The 88in and 109in Land Rovers were left alone, and it was during this period that overseas markets were steadily lost to Japanese competitors. For many customers the Land Rover was not only becoming outdated, but also poor value-for-money compared to newer offerings. Models such as the Toyota Land Cruiser and Nissan Patrol were proving very reliable vehicles with excellent service back-up, something that was being abandoned in many markets under British Leyland control.

Indeed, within the whole British Leyland empire there was increasing turmoil. Productivity and profits had slumped and it became clear that without government support the conglomerate would fail, with the loss of thousands of jobs. The Labour government of the time appointed a team to conduct an assessment of the company, with an initial report due in March 1975.

The team was led by Sir Don Ryder, head of the newly formed National Enterprise Board. The team concluded that there needed to be an extensive programme of rationalisation both in products and manufacturing facilities, plus an improvement in standards of build quality and reliability. The Ryder report also proposed new government loans and equity capital totalling £700m, which meant the company would now become nationalised and under the control of the National Enterprise Board. It was renamed British Leyland Limited, and broken up into four profit centres, with Rover-Triumph being amalgamated under the BL Cars group.

After two years, the company was reorganised again. On 1 November 1977 Michael Edwardes was appointed to run BL Limited. Investment for Land Rover had actually been reduced since 1975, but now the tide began to turn. Following Edwardes' arrival, the 'Leyland Cars' one-badge-fits-all policy was reversed, and thoughts began to form to separate Land Rover from the volume cars group.

Land Rover Limited was created as a separate operating division in July 1978, marking the first time Land Rover had been treated as an independent company from its Rover or BL parent. This marked the beginning of some real investment in the company – and with it came some long-awaited new product development.

David Bache era and Range Rover

The 6x4 Range Rover fire tender was developed by Carmichael. (Author's collection)

Land Rover Design – 70 years of success

Following customer feedback, the interior was given several upgrades over the next couple of years to include a non-slip mat in the trunk to avoid gun dogs slipping around, thick pile carpet on the centre tunnel and a moulded brushed nylon headlining. In this example seats have been re-trimmed using the Bri Nylon fabric available as an option from 1973. The clock only was fitted as standard, the other minor gauges were optional. The gaiter on the Hi-Low gearlever and handbrake is missing here.

Geof Miller and Roger Crathorne stayed with the Range Rover project as a forward development team into 1972. This Bahama Gold prototype – seen here near the old Solihull test track – was an early attempt at a four-door by Land Rover, and completed in late 1971, but was not progressed further. According to Crathorne it was later scrapped and the vehicle rebodied as a two-door. Note the slightly bigger rear door fixed glass pane compared to the eventual production model. (Courtesy Roger Crathorne)

Land Rover paint colours

The Series Land Rover colour palette comprised just six colours, used extensively from the Series II up to the end of Series III production in 1985. They were essentially functional pastel colours, based around usage and location. They were also deliberately 'chalky' in composition, with a lot of white in the formulation, which meant they still looked good when left unpolished for long periods.

Light green had been used as a generic colour during WW2 for agricultural machines and buildings. It was also a typical primer colour for aluminium components, particularly in the aircraft industry. However, the idea that the original pale green colour was employed to use up old paint stocks is a complete myth – it was a standard Rover P3 paint finish.

The popular Bronze Green colour was a reformulated version of a standard British Army colour, and replaced Light Green as the only available option from June 1949. Mid Grey was another regular agricultural colour choice of the 1950s, used on Ferguson tractors and many commercial vehicles of the period, and was thus easy for farmers or hauliers to repair or touch-up.

Limestone started out as the contrast roof colour for Station Wagons and Hard Tops from 1954, but later became a regular body colour. From Series IIA onwards it became the standard wheel colour too for all paint finishes except Bronze Green. It proved a useful choice for commercials or emergency vehicles to allow for signwriting as it did not show the dirt and dust as badly as a true white. Sand was another pale colour useful for camouflage in deserts or for tropical climates to keep the car cool, while Marine Blue reminded many people of a typical Cotswold blue that was a popular 1950s car colour. It also seemed to fit in perfectly with the English landscape and, together with Bronze Green, was the most popular choice for British farmers.

The Range Rover had a six colour line-up of Masai Red, Bahama Gold, Lincoln Green, Tuscan Blue, Davos White and Sahara Dust, all colour developments of Rover car paints of the period.

The introduction of TPO paint and expansion of the paint shop in 1979 allowed a wider range of colours for all models, including some bright colours for the Land Rover for the first time. These included four Triumph TR7 colours: Java Green, Pageant Blue, Inca Yellow and Russet Brown, initially available only on the Stage 1 V8. According to Roger Crathorne, these new colours came about due to his recommendations following a trip to Costa Rica that customers wanted livelier colours, not drab ones from the UK.

After the launch of Defender in the 1990s, the decision was taken to promote the vehicle with regular updates to colours, wheels and options as a way to promote sales, much as in the volume car industry. This was in stark contrast to the previous approach, where Sales and Marketing had tended to focus on new functional features to generate sales, features that had been developed by Special Projects in response to a demand from specialist clients.

The paint shop switched to using clear-over-base (COB) paint finishes for all Lode Lane production in 1993, in preparation for the new P38A Range Rover launch. Thus full gloss and metallic paint finishes were available on the Defender, including four special micatallic colours for County models: County Black, County Red, County Grey and County Green.

The introduction of the classic Land Rover colours in the 1990s coincided with the general retro design trend that flourished at that time, with a sentimental yearning to rediscover past icons such as the Mini and VW Beetle. The 1999 'Heritage' special edition reintroduced Bronze Green and a reformulated light green solid colour, called Atlantic Green. The NAS limited edition 'LE' Station Wagons of 1997 were painted in another version of the original light green – Willow Green – with a contrast white roof.

The classic six colour line-up used for Series Land Rover models barely changed in 25 years. (Author's collection)

The original Range Rover colour palette was deliberately closer to Rover car colours of the time, particularly the use of a deep red, blue and mustard yellow. (Author's collection)

1978-1989

Chapter 3

Independence and expansion of the range

The 1975 Ryder Report had recommended that a new integrated BL Engineering Division be formed, based at Monkspath near Solihull, and to be headed by Spen King. Although buildings were found and initial concept work was started – including architect's plans – the proposal was slow to gain momentum, not least due to the in-fighting of the various BL engineering offices around the Midlands.

Following Michael Edwardes' arrival the Engineering Division proposals were torn up and a smaller research operation called BL Technology was created, with Spen King once again at the helm, based at a new location at Gaydon, south of Warwick.

RAF Gaydon had been built during WW2 and had been developed into one of the country's main V-Bomber bases during the 1960s, with a very long runway. Now the site was no longer needed by the RAF it was offered to British Leyland to provide a vehicle testing facility and proving ground, incorporating the 1¾-mile runway as a high speed track. This was a facility the company had sorely lacked, and BL had to rely on the aging MIRA facility at Nuneaton for the majority of vehicle proving. This was in marked contrast to other companies. Vauxhall had built a fabulous facility at Millbrook, and Ford had developed joint facilities at Lommel in Belgium and Boreham in Essex. Thus, BL Technology was created in 1979 and, for now, Engineering and Styling would remain within the build plants. It would be another 17 years before a group-wide Engineering Centre would come to be based at Gaydon.

Edwardes was fairly satisfied by what he found at Solihull, and was keen to expand production capacity. As part of his reorganisation, £280m of new government funding was secured. This was to be the start of major investment in production capacity and new model development that was hoped would lead to big increases in sales into the 1980s (see sidebar opposite).

This independent Land Rover company was headed up by Mike Hodgkinson, an ex-Ford finance manager who had joined BL in the early 1970s. Hodgkinson inherited a company manufacturing 55,000 4x4 vehicles per annum, but with just one product under development. This was a replacement for the 88 and 109in Land Rover. Although the styling was virtually the same as the existing vehicles, there was a new chassis and many changes under the skin. Three versions of the new Land Rover were to be available in 90, 100 and 110in wheelbases, with two new engine programmes also under way: a five-bearing petrol engine and a V8 installation, introduced within a modified Series 3 Land Rover, known as 'Stage 1.'

At first sight, Hodgkinson's new Land Rover company looked appealing: a small workforce, an apparently profitable range of vehicles and – finally – some new programmes under way. Military vehicles formed a big part of this operation, with several versions of the Series II and III being produced, plus the military 101in V8 Forward Control version.

The Stage 1 V8 Land Rover was developed on the 109in chassis to offer a high-performance version, to open up Land Rover's market

Independence and expansion of the range

Land Rover Stage 1 and Stage 2 reorganisation

With £280m of new government funding announced in August 1978, Land Rover Limited was able to start major investment in new model development and production capacity to take Land Rover into the 1980s. The plans were split into 'Stage 1' and 'Stage 2' phases.

Land Rover and Range Rover production had always been constrained by available capacity at the various satellite plants used to build sub-assemblies. V8 engine build at Acocks Green was the biggest bottleneck, with a capacity of just 850 engines per week.

Stage 1 was an initial investment of £15m to cover the expansion of V8 engine capacity, to allow 2000 engines per week to be available across the BL Corporation, including the Rover SD1 and Triumph TR8 models – both built at Lode Lane. Concurrently, there was expansion of the Range Rover gearbox and central transmissions facility. This allowed the launch of the 'Stage 1' V8 109in Land Rover in 1979, with its detuned version of the Range Rover V8 engine installed in the 109in Land Rover chassis. As part of this Stage 1 phase, the empty North Works (where P6 had been built until December 1975) was converted to become an automated factory parts and packing depot, with the mothballed P6 paint shop brought back into use to paint both Land Rover and Range Rover bodies.

The £250m Stage 2 investment saw the North Works then converted to include engine machining and assembly of engines, initially the four-cylinder petrol and diesel, later the V8. Engine build at Acocks Green was wound down and the plant sold off. Next up was the Range Rover build. A further £15m was invested to expand the North Works for a new Range Rover production capacity of up to 600 per week, particularly the new four-door version. This now allowed Land Rover production to expand to fill all of the original South Works, from two assembly lines to seven lines. This facilitated the build of more County models and the High Capacity Pick Up body.

From January 1982, production of the SD1 saloon would be transferred to Cowley plant in Oxford and the TR7/8 build at Solihull would be stopped. Thus for the first time Lode Lane became a purely Land Rover production plant, operated by Land Rover Limited. The redundant East Works plant was touted around the motor industry for sale, but there were no takers, and – remarkably – the facility lay vacant for the next two years.

With the arrival of Tony Gilroy in January 1983, the company was again reorganised. Seven satellite plants around Birmingham were closed down, with a 16 per cent reduction in workforce to 9700. Production operations were now centred at Lode Lane, ending the scattered pattern of Land Rover assembly over multiple sites in Birmingham that had existed over the past two decades. Pengam in Cardiff remained for car gearbox assembly, but would also later close down. To make way for the relocated sub-assembly operations, other non-production teams such as Service, Marketing, and the development workshops were moved off-plant to various sites around Birmingham or Coventry. With the East Works lying dormant, Gilroy wasted no time in putting forward plans to utilise the plant as a press shop, BIW assembly and transmission manufacturing area.

In March 1983, the 'Stage 2' programme reached fruition with the launch of the Land Rover One Ten with its new coil spring chassis. 'Stage 2' also included the 127 model with extended load bed, available as a Crew Cab or Chassis Cab configuration.

The final part of 'Stage 2' was the Ninety SWB launched in June 1984. Land Rover production capacity out of Lode Lane was now up to 2700 per week, more than double that available in 1978.

Lode Lane plant, circa 1989, looking north. The P6 North Works was doubled in area to allow Range Rover assembly. In the centre is the paint shop, linked by overhead conveyors to the £30m East Works (right). This was completed in 1976 to build SD1 but was mothballed by 1982. (Roger Crathorne)

share in Middle East markets. Launched in 1979, it used a detuned 91bhp 3.5-litre V8 engine with the complete LT95 gearbox, transfer gearing and permanent 4WD from the Range Rover. To accommodate the installation, the V8 engine was mounted further forward, and this, together with the need for a larger radiator, required a new front end restyle with the grille pushed forward flush with the fenders – a style that endured until the final outgoing Defender model in 2016.

A further development of the 109in chassis was the High Capacity Pick Up, launched in 1982. This had a large aluminium rear body tray that was 8in longer and 1.7in wider than standard to offer a 2-metre (79.2in) load bed length and 1200kg payload. Crucially, this allowed it to compete with the Toyota Land Cruiser Pick Up in export markets.

That same year, Tony Poole's team of three designers responded to requests from Marketing to come up with a package of improvements to the basic Land Rover Station Wagon, to appeal to the burgeoning 4x4 leisure market into which Japanese brands such as Toyota, Nissan and Suzuki had made such inroads. The resulting 'County' models were jazzed-up with silver bezels on the headlamps, a pair of Lucas fog lamps, tinted glass, and a spare wheel cover on the tailgate. 88in models featured a broad four-stripe decal down the sides, while 109in models received a more tasteful two-stripe design. Additional features included a new comfort seat trimmed in grey houndstooth tweed with rake adjustment and headrest, together with a lidded centre cubby box for oddments.

The increased investment certainly eased the production constraints, but sales had hardly improved and there was an air of disappointment within BL that the money had not delivered the expected results. Worse still, Land Rover was losing money badly. In January 1983 Hodgkinson was replaced and Tony Gilroy was appointed as Managing Director. Like Hodgkinson, Gilroy had started his career with Ford, in

Land Rover Design – 70 years of success

The Stage 1 V8 Land Rover arrived in 1979 and was the first model to use the extended bonnet panel. It was also offered in brighter Triumph TR7 colours.

Early One Ten sketch in marker and pastel by Paul Taylor, 1980. (Courtesy Paul Taylor)

Independence and expansion of the range

Four Stage 2 hybrid prototypes were built in autumn 1976 by Roger Crathorne's team in Vehicle Development with wheelbases of 90-, 100- and 110in. This 100in hybrid was built as a four-door Station Wagon with canvas rear tilt. It survives in the Dunsfold Collection in Surrey. (Courtesy Dunsfold Collection)

1954, and had previously run Freight Rover's van division. A tough, no-nonsense Irishman, Gilroy immediately set about implementing a drastic cost-cutting and rationalisation plan that resulted in the closure of satellite plants around Birmingham and a reduction in workforce (see sidebar, page 67).

Two months later the Land Rover One Ten model was introduced. This had started out in 1976 with a series of four Range Rover chassis being fitted with Land Rover body sets. Two were left with an unchanged 100in wheelbase, one was shortened by 10in and one was given a 10in stretch in wheelbase. The permanent 4x4 drivetrain and V8 engine were left largely alone and the suspension used long travel coil springs all round. The body width was unchanged and the wider track of the Range Rover was covered using simple add-on wheelarch extensions made of plastic.

Following successful trials at Eastnor Castle, the 'Stage 2' programme was signed off for development in April 1977. Very early into the new programme, it was decided to drop the 100in version and concentrate on the 110in Stage 2. To cover the loss of the 100in

Land Rover Design – 70 years of success

The introduction of the Ninety saw a final change to a one-piece door and a new push button door handle design. Note the lower fuel filler position for the side-mounted tank. The standard rear tilt colour was Khaki, seats were black vinyl.

it was decided to continue the build of Series III and continue the lucrative market in Africa for the Series III CKD versions.

This decision was to lead to severe design compromises, due to the number of parts which were required to be carried over between Series III and Stage 2. Arguments raged over whether a proposed one-piece windscreen was really necessary for all versions and why could not the two-piece screen be maintained. New doors with drop down glasses were designed but not implemented as the separate door upper frames were still felt to be important for certain markets. And so forth.

Such petty rows and interminable delays were symptomatic of the haphazard way that the Engineering Department was run under Tom Barton and by 1980 he had retired, to be replaced eventually by Bill Morris as Engineering Director. Admittedly resources were scarce but nevertheless, taking a full seven years to produce a revamped vehicle as simple as the Stage 2 Land Rover was desperately slow even by 1970s standards.

Although initially based on Range Rover architecture, the production One Ten employed a new, much stronger, chassis with deeper side members and bigger coil springs. Inevitably, it utilised much of the existing Series III body in all the existing variants: Soft Top, Truck Cab, High Capacity Pick Up or Station Wagon. Engines were the existing 2.3-litre four-cylinder units in either petrol or diesel version, or the 3.5-litre V8. The front end used the long bonnet from the V8 on all versions with a new eight-bar slatted black grille, flanked by black plastic trim finishers around the headlamps, a style that defined the look up until the end of Defender production. To meet new European vision regulations, a deeper one-piece windscreen was fitted and the old double-skinned tropical roof was dropped in favour of a single skin type, with extra insulation provided by a new internal headliner.

The interior was revised with a new instrument binnacle and four-spoke steering wheel, houndstooth tweed seats, an improved heater and provision for a radio/cassette. At the One Ten launch, Land Rover also showed a prototype Double Cab pick-up version, which was offered for sale one year later. This 'Land Rover 127' had a longer 127in wheelbase utilising rear doors from the Station Wagon and a cropped rear tray from the High Capacity Pick Up to give a useful Crew Cab model, fitted with the V8 engine.

Despite the Stage 2 investment, sales of the new One Ten stayed well below par throughout 1983, with volumes of just 200 a week compared to over 600 of the old Series III that was still in production. At this point, 60 per cent of Land Rover production was long-wheelbase – hence the priority for the One Ten model.

To combat the dramatic fall-off in sales, an intense course of new model development was started. In the short term, the Stage 2 short wheelbase introduction was brought forward to summer 1984. This used a new configuration with a 92.9in wheelbase, although it was marketed as the 'Ninety' for convenience.

Compared to the One Ten, the Ninety was offered in all the same basic configurations but the engine line-up at launch was four-cylinder only, with no V8 option.

One year later the V8 engine was also offered in the short wheelbase model, together with a change to a more powerful 2.5-litre four-cylinder engine, available as an 80bhp petrol or 65.5bhp diesel. At this point, the Series III finally ceased production at Lode Lane, with over 427,000 having been produced since 1971.

> **Rover studio developments in the 1980s**
>
> In the early 1980s, BL had a number of styling studios around the Midlands. Austin had its studio at Longbridge plant, led by Rex Fleming and Harris Mann, which was based in two separate buildings. The first of these was called the Elephant House, a circular building built originally as a commercial vehicles showroom in 1965. The facility was hardly ideal: the central area could be adapted for displays and presentations while designers and modellers worked around the perimeter, with makeshift screening on the windows to avoid passers-by seeing in. In true Longbridge Brummie humour, it was dubbed the 'Elephant House' as a comic reference to the pachyderm house at nearby Dudley Zoo.[1]
>
> The second building was known as the 'Glass House,' a facility built in the mid-1950s for styling and prototype build. It had long, sloping frosted windows, wooden parquet floors, and had received little in the way of updates over the past decades. There was no dedicated space to view models outside within the busy factory complex.
>
> Rover, of course, had the upstairs facility at Lode Lane led by Bache. Since 1973, Bache had also assumed responsibility for the Triumph studio at Canley in Coventry, and split his time between the two studios. Finally, there was the small Jaguar styling studio at Browns Lane managed by Doug Thorpe, but this always remained very independent of the other studios, and did not engage with them. Indeed, there was little fraternisation among any of the different BL studios, and transferring jobs between them was positively discouraged. Most significantly, all of the design studio chiefs were under the control of Chief Engineers – in Land Rover's case, Bill Morris.
>
> After Bache was dismissed in late 1981, Roy Axe came in to assume responsibility for Austin-Rover, but was shocked by what he found. "You could say that I should have looked before I joined, but it's very difficult to do that with design departments; you can't be given a tour before joining the company for obvious reasons. Equipment was best described as rudimentary, and there was no showroom at Canley. It was all in poor shape and quite awful. I was really stunned and really wondered what I had got myself into."
>
> At first, Axe based himself at Canley, but kept the Longbridge studio to continue existing projects such as the Maestro, Montego and Ambassador. However, within months Harris Mann, Rob Owen and Roger Tucker were out of the picture. In their place, Axe promoted Gordon Sked to head up exterior design, and newly arrived Richard Hamblin for interior design.
>
> Land Rover was in the middle of the Stage 2 developments, and as part of the massive reorganisation of Lode Lane the small design team moved offsite to a small facility in Drayton Road in early 1981, headed up by Tony Poole.
>
> It was not just facilities that Axe lacked, it was staff. Spring 1983 saw the closure of Chrysler-Talbot's design studio at Whitley just down the road from Canley, which led to a sudden glut of designers looking for new positions around Coventry, and Axe was suddenly receiving CVs and portfolios from former colleagues on a weekly basis. Wasting no time, he swiftly recruited Gerry McGovern, David Arbuckle and Dave Saddington to head up the Rover and MG teams in Canley, plus a number of skilled clay modellers.
>
> *1. In fact, the Dudley Zoo building is rectangular and part of a fine series of 1930s Art Deco buildings by the Tecton Group, who also built the penguin pool at London Zoo. The reference probably stems from a combination of its modernist style, circular plan form and stepped roof structure that was vaguely reminiscent of zoo architecture. In addition, in 1964 London Zoo had also opened a new Elephant House, designed by Sir Hugh Casson, which was also circular in form but architecturally quite different.*

Land Rover Design – 70 years of success

Roy Axe replaces David Bache

It was not just the senior directors of BL Limited that were undergoing rapid change. Within the design studios there was upheaval, too. David Bache had steadily risen in the ranks to become the overall design chief within the BL Empire, and had taken responsibility for all of Triumph and Rover passenger car design activity, plus of course Land Rover. Harris Mann was still in charge of the Austin-Morris exterior studio at Longbridge but was under increasing pressure from Bache, who was keen to push his own ideas for the forthcoming LC1 and ADO 88 programmes for Austin-Morris.

However, Bache began to overreach himself, and assumed an affected air that sat uncomfortably with the hard-nosed executives now running BL, not least Chairman Harold Musgrove. In autumn 1981 matters came to a head when, during a crucial sign-off meeting for the Austin Montego, Bache was seen to waffle his way through a key presentation and failed to convince the assembled executives about the multiple concerns with the design, particularly the issues on manufacturing quality. As the discussion became more heated and Bache constantly interrupted the Chairman, he was told multiple times to 'shut up,' then finally ordered out of the room. The following day Musgrove fired him unceremoniously.

It was a sad end to what had been a glorious career at Rover but it seems Bache had failed to appreciate how deadly serious the whole product development process was becoming, with Board members having little patience for whimsical styling presentations in place of coherent design strategies that could satisfy them about a new model's chance of success against its competitors.

Bache's PA, Maureen Hill, was able to see the situation close up. "Harold didn't like to be put down in any shape or form. Because David's brother was Sales Director at Austin, David had gone into Austin as an Engineering apprentice which was a higher scale than Harold as a Works apprentice. So from day one they had a problem. It sounds daft but that's how petty it was."

Chrysler designer Roy Axe was discreetly approached to see if he would be willing to take on the responsibility of overall Design Director for Austin-Rover. The initial approach was made by Harry Sheron, an old colleague from Axe's Chrysler days who was now in charge of operations at BL Technology with Spen King. On meeting Musgrove, the situation at BL was explained to Axe, what was wrong, what was right and what he wanted to put right. "[Musgrove] told me that the company was in very fine shape regarding manufacturing," commented Axe.[1] "They had this confrontation with the workers but the manufacturing division was going to have a lot of money poured into it. Manufacturing was going to be computerised, and robotised at the plants, and new models were in the pipeline. However, he was not satisfied overall, and he felt the weakness in the operation was styling."

Axe duly took over in January 1982 and instantly set about reorganising the design activities, relocating the main car studios from Longbridge and Solihull to a new base at Canley in Coventry. In fact, this initiative had been started by Bache, but was given new urgency under Axe's arrival. Axe's priorities were focused on the urgent concerns with the imminent Austin Maestro and Montego launches, not to mention the need for a raft of new model programmes using Honda technology. The main Rover car team were transferred over to Canley to start work on a major project with Honda – the XX – to be launched in 1986 as the Rover 800.

A new start for Land Rover Design

It must be emphasised that Land Rover was still a totally separate division within BL, and reported quite independently up to the main BL board based in London. Axe's remit did not cover Land Rover, so had no immediate impact. However, Land Rover design was also going through a step change.

The Stage 1 and Stage 2 Land Rovers were very much the creations of Engineering and Special Products, with only limited involvement from the styling studio. Minor areas such the instrument binnacle and seats would be mocked up, but there was scarcely any need for elaborate clay models. Under the Stage 2 developments at Lode Lane, the small Land Rover design group moved off-site in 1981 to a facility in Drayton Road, Solihull, headed up by Tony Poole, with Vic Hammond and Mehmet 'Memo' Ozozturk as stylists, and four modellers: Phil Scrivens, Mick Jones, Peter Fox and Chris Stevens. It was an arrangement that suited the old-school engineers such as Tom Barton, who had little time for the 'felt tip boys' in Styling, regarding the task as being something that his engineers alone should control.

"Land Rover came from an ethos of hose-out interiors and operating switches with gloved hands, the engineers had instilled that with us," recalls Jones. "They weren't too keen when stylists came in with their fancy ideas …"

The facilities were very basic. Heating was poor – a challenge with clay models – the gable end wall moved when it was windy and squirrels lived in the roof. "But it was our studio, our opportunity to improve, and within a year we were renamed Advanced Design Operations," he continues.

Gilroy's ambitious new model plans meant that design activities needed an overhaul, with more manpower and a far more professional team as they went into the 1980s. To lead this team, a new design chief was brought in – David Evans. Evans had studied at the RCA and joined Axe's studio in Whitley in 1973, so there were old links here. With the disbanding of that studio ten years later, he was casting around for a new position when the opportunity at Land Rover cropped up. Evans joined the Drayton Lane studio in October 1983, initially as Poole's assistant, although within three months he was appointed Chief Designer.

The first priority was various Range Rover updates and 'project Capricorn' – an update of the Ninety to delete the centre bulkhead

---

1. From his autobiography – A Life in Style.

Independence and expansion of the range

Capricorn was intended to have a totally new IP. (Courtesy David Evans)

Capricorn also featured numerous revisions to modernise the Land Rover for the 1980s. (Courtesy David Evans)

to accept forward-facing rear seats, add a new IP, and face-lift the exterior. To support the work, staff headcount was increased to 15, including George Thomson and Dick Bartlam as stylists, while to lead the modelling team Evans recruited Matt Muncaster, an old colleague from Whitley, who duly brought along three skilled clay modellers: Tom Mosey, Gordon Henderson, and Phil Randle.

Both Evans and Thomson credit Matt Muncaster with a lot of the early success of the team. "In our early days we could have been frustrated in our work without Matt's 'wheeling and dealing' to get us what was required to do the job," says Thomson. "His knowledge of modelling and modellers was an invaluable part of the team's success." Nevertheless, 'Capricorn' was deemed too difficult for the time, and it would be another 20 years until forward-facing seats and a new IP would be incorporated into the Defender.

Evans reported to Engineering Director Bill Morris, whose attitude to design was an improvement on Tom Barton's. "Bill understood fully the value of Design and gave tremendous support throughout his tenure without ever interfering with the visual elements of the design," recalls Evans. "Gilroy could be different, though, he was certainly not a man to be trifled with ..."

As the team increased, it became obvious that a dedicated colour and trim group was needed, and Mick Jones was delegated to lead it, looking after paint colour development and new fabric trim for the County models. New systems and processes of colour control were needed. "During early negotiations with Rover Cars' team we discovered many links – references, names and colour matches – remained between Jaguar-Rover-Triumph colour ranges that Land Rover needed to honour," explains Jones. "During this time we supported colour and trim design reviews with body paint 'take rates'[2] and switched emphasis from car-based, solid, flat colours into a more appropriate complementary range, and developed new technologies with the full assistance of the laboratories and the production paintshop."

False starts: Llama, Ibex and Inca

Work also commenced on a new off-road leisure model based on the Range Rover chassis, leading to 'project Jay,' covered later in this chapter, and early studies for a completely new Range Rover – 'project Pegasus' – covered in the next chapter.

In the shorter term, a new military project codenamed 'Llama' was begun in 1984. This was for a new Forward Control light truck based on the One Ten, that might offer better civilian sales potential than the 101in V8 Forward Control model. George Thomson led the design, coming up with a simple yet harmonious truck cab that kept a Land Rover face with round headlamps and horizontal grille bars.

---

2. *Popularity of each colour by sales volume.*

Land Rover Design – 70 years of success

The tilting cab used a steel and aluminium frame with GRP panels made by Reliant, but there were some doubts within the military as to its durability. The Llama used heavy-duty Salisbury axles with 8in wider tracks than the standard One Ten and offered a generous cargo bed that could be fitted with a dropside body or a fixed side with canvas tilt. It retained the 135bhp V8 drivetrain with the LT77 5-speed manual gearbox and permanent 4WD and one prototype was fitted with the underpowered 2.5-litre turbo diesel engine. However, the Army wanted an all-diesel fleet. Evans explains the situation: "Llama was ill-fated because the MoD wanted an automatic diesel vehicle, but were offered up a petrol manual one only. It wasn't a design failure." With no military support, the Llama could not justify sufficient civilian sales to go ahead, and, after 11 prototypes were built, the project was abandoned in 1987.

Llama clay model in Drayton Road studio. Note the Norton rails used to aid modelling. (Courtesy David Evans)

Independence and expansion of the range

Mock-up of the Llama cab. In typical Land Rover practice, the IP was non-handed and the instrument pod could be swapped to the passenger side for LHD. (Courtesy David Evans)

Then there was 'Project Ibex/Inca.' This was for a new Land Rover and Range Rover based off a single common structure. "We evaluated it but the compromises proved too great, especially regarding the [greater] tumblehome required for Range Rover," says Evans. The idea was soon dropped.

Lastly, there was an attempt to rejuvenate the Land Rover Ninety as a soft top leisure market 4x4 to capitalise on the market opened up by the cheaper Suzuki SJ and Daihatsu Rocky ('Fourtrak' in UK). The 'Cariba' concept was designed by Mick Jones and Alan Sheppard in 1987, based on a Ninety V8 with a roll hoop linked to the front header, spoked 205-16in steel wheels, and a side-opening rear tailgate with spare wheel carrier. The doors were cut down at the belt line, a prominent bull bar and spotlamps were fitted at the front, and the package was completed with a metallic paint finish and seats trimmed in leather.

Two examples were built by SVO and shown to the press, but the project did not proceed further. The design team was busy on the 'Jay' and 'Pegasus' projects and the sales side was fully occupied on the US launch of the Range Rover.

Four Llama prototypes are housed at the Dunsfold Collection. This is prototype No 1. The rectangular notch cut into the grille is so that the cab could clear a front-mounted towing hitch when tilted forward. (Courtesy Dunsfold Collection)

Land Rover Design – 70 years of success

Drayton Road and Block 38A studios

Although Tony Poole had handled Land Rover design for the previous two decades, by now he was suffering health issues that hastened the establishing of a new team. This would be staffed by university-educated designers, with a deeper set of skills than Poole's small group of six could offer. In order to progress Tony Gilroy's ambitious model plans, a new Chief Designer was brought in to reinvigorate Land Rover design.

David Evans arrived in October 1983 to run what was termed 'Advanced Design Operations' at Drayton Road, Solihull, although it encompassed all of Land Rover studio activities. This was an off-site workshop that had been set up in the wake of the reorganisation of Lode Lane in January 1981, when Bache's old Rover studio was being closed and the cars team moved to Canley. The name was a step up from the term 'styling' at least, but reflects the deeply held belief by engineering bosses within British Leyland not to yield the term 'design' to a creative studio activity, believing that 'design' and 'designers' were terms that belonged to engineers within the company. Thus, 'Advanced Design' was attributed to styling operations.

"I didn't see the design department until I arrived, they deliberately didn't want me to! It was past a car valeting depot, you got a bit damp on the way in," explains Evans. The facility included development workshops and the press cars department and had around 200 staff on site. The design area comprised two parts, one for designers and a larger area where the models were done. Apart from Poole there was Mehmet 'Memo' Ozozturk, three modellers including Mick Jones, and designer Mike Sampson, who joined from Jaguar in 1983.

In May 1984, Evans recruited George Thomson and Dick Bartlam to assist him on the design side, together with Matt Muncaster as modelling manager – all former colleagues from Chrysler's Whitley studio. Mick Jones was delegated to set up a Land Rover colour and trim section, something that had never really existed before.

"Drayton Road studio was one end of a workshop," continues Evans. "Matt Muncaster bought a set of Norton Rails from Whitley, we used it for the Llama project. Prior to that there were no accurate rails to take measurements on the models." Project Jay was the next big programme that formed the bulk of the work at Drayton Road for the next two years, with a series of full-size models being produced there. "In fact, we started thinking on the new Range Rover from early 1985 as sketches, then full-size tape drawings – we had a vast area where we could pin up work. Then the Range Rover programme stopped because Jay was more urgent."

In August 1986, the decision was taken to move back to Lode Lane into a newly refurbished facility. This was housed in one half of Block 38 (38A) at Lode Lane, right next to the 'jungle track' in Billsmore Wood. When set up, the studio's main function was the new Range Rover project, which henceforth became known as 'Project 38A,' or P38A for short. Shortly afterwards Tony Poole retired after 30 years of involvement in Land Rover design, the last of the original Bache team from the 1950s.

By 1988, David Evans' design team had expanded to more than 60 staff. New designers included Don Wyatt, Alan Mobberley, Mike Brogan, Alan Sheppard, Peter Crowley, David Brisbourne and Belinda Dolan. The colour and trim group under Mick Jones now comprised John Stark, Karen Hind, Kim Turner and Gary Cox. On the modelling side there was a team of 14 modellers under Matt Muncaster, and 11 feasibility engineers including Pete Ludford, Peter James and Clive Jones.

"38A was a nice little studio, well equipped," recalls Mick Jones. "The walls were white, the floor covered in yellow thermo-plastic tiles. We had row upon row of colour-corrected fluorescent lighting, thermostat-controlled heating, a large turntable, model plates and Stiefelmayer 3D measuring equipment." Outside there was a viewing yard with gates on either end because the main South Block perimeter road ran through it. "When we had a viewing we had to close the gates, stop the traffic going by and upset everyone ..." says Don Wyatt.

"Ah, but we could grow tomatoes in the glass annexe at Drayton Road!" adds Mike Sampson.

Inside Drayton Road facility, not long before the studio was closed. This shows George Thomson's side of the Jay model in comparison with Range Rover. The impression of a shorter rear overhang is clearly evident here, although the actual chassis is unchanged. (Courtesy Mike Sampson)

Independence and expansion of the range

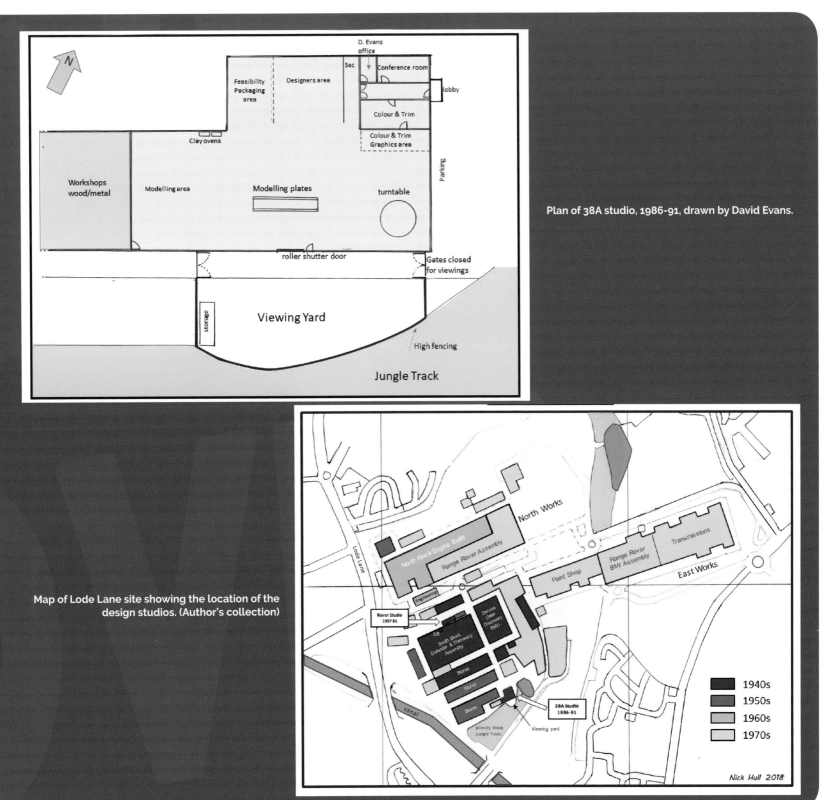

Plan of 38A studio, 1986-91, drawn by David Evans.

Map of Lode Lane site showing the location of the design studios. (Author's collection)

Land Rover Design – 70 years of success

The design team at 38A studio, for the occasion of Tony Poole's retirement, 1987. Bill Morris and David Evans are on the left, Poole is in the dark suit. By this stage the total team was over 60 strong. (Courtesy Mike Sampson)

Range Rover moves upmarket

If the Land Rover had suffered from a lack of development, so too had the Range Rover. Since its launch in 1970 it had received little funding, not least because of the severe production constraints and continued high demand for the vehicle around the world. 'Why bother changing a winning formula when we can sell every one we build' seemed to be the mantra of BL during the 1970s. However, by the late 1970s sales were dropping sharply and Land Rover's complacency needed to be addressed. A number of specialists had developed niche versions of the basic two-door Range Rover including four-door bodies, long-wheelbase variants, automatic transmissions and diesel engines. Within Land Rover Engineering there appeared to be a degree of derision towards these outside efforts, and serious developments continued to be handled in a somewhat piecemeal fashion.

Land Rover Special Products prepared three prototypes that would pave the way for full production versions if they proved successful. In something of a change of heart, these models were developed with the assistance of outside specialists, but only so as to minimise Engineering's own expenditure on resources and act as an insurance against failure. March 1980 saw the introduction of the Monteverdi four-door conversion and – although Land Rover had previously built a four-door prototype Range Rover – it was the Monteverdi conversion that it promoted. Despite the truncated rear passenger doors compared with the final product, the overall view was that this Swiss conversion was pretty slick and well executed.

Special Products approved the car for production and offered it for sale through Land Rover dealerships. Of course, Land Rover cannot have been encouraged by the Monteverdi's pitiful sales of around 130 units (it was an expensive conversion) but the reaction

Independence and expansion of the range

Monteverdi Range Rover was offered at £16,507 in the UK – a hefty 33 per cent increase over the standard two-door. Note the unique rear door shut line around the wheelarch.

The 'In Vogue' special edition two-door was developed in February 1981, in conjunction with Wood and Pickett. The production run was announced as 1000, with around 250 examples being sold in the UK at £13,800.

to the four-door concept gave a renewed impetus to introduce an in-house version.

Land Rover revisited its own four-door Range Rover development. The limited engineering resources were continually stretched with work on Stage 1 and Stage 2 programmes and Hodgkinson became ever-more frustrated at the lack of progress on any Range Rover work. After endless delays he commissioned Carbodies in Coventry for some of the development work, and the in-house version of

Land Rover Design – 70 years of success

In Vogue four-door, 1982, photographed at the Burton Dassett hills, near Gaydon. Within a couple of years over 90 per cent of Range Rovers were four-door models and the two-door was dropped from the UK market in 1984. The three-spoke alloy wheel first seen on the In Vogue prototype was known as the 'Mustang wheel' within Styling.

the four-door Range Rover was announced in July 1981.[3] Changes over the two-door were minimal but the rear seat was moved 3in rearward to aid egress and provide more legroom. According to Mick Jones, the initial design work was hampered when it was discovered that the original two-door rear fenders were different lengths! The four-door was built on a new line in the revamped North Works, in an extension to the east, and the new paint shop facilities allowed a wider range of colours – including metallic paints – for the first time.

The second project was to offer an automatic transmission option, something the Middle Eastern markets had been requesting for years. In late 1980, Schuler automatic transmission conversions were offered, using the Chrysler Torqueflite 3-speed transmission incorporating a chain-driven transfer box and ABS. It was another expensive conversion at £2300 plus VAT, but over 200 examples were produced, and the success of the project convinced Land Rover to push ahead with its own automatic version, still using the Chrysler gearbox, which went on sale in summer 1982.

The third initiative came about via a photo-shoot with fashion magazine Vogue in which a Range Rover was used as a background prop for the Jaeger collection. The Land Rover designers worked with London coachbuilder Wood and Pickett to prepare a luxury two-door version painted in pale blue with a picnic hamper in the fully-carpeted trunk. This led to a limited edition 'In Vogue' model being offered by Land Rover in February 1981 at £800 over the basic price which replicated the photo-shoot vehicle. The specification featured air-conditioning as standard, walnut door cappings, and a storage console between the seats, and was finished off with a picnic hamper and special Range Rover cool box. The vehicles were finished in Vogue Blue metallic, and thanks to the success (and sizeable profit) of the 1000-off limited edition, a second series of 500 was produced in 1983, and the In Vogue became a production model in its own right from 1984.

The 'In Vogue' special edition was the first attempt by Land Rover to deliberately position the Range Rover as a luxury fashion statement and to create more overt product placement opportunities with other luxury brands, particularly British ones such as Simpsons of Piccadilly, Daks or Jaeger. Today, this type of product association is a standard part of premium brand marketing, but at that point it had not been really capitalised. Marketing at Land Rover had traditionally focused on engaging Special Products to develop versions that added some specific function to the vehicle – a new implement, a longer chassis, some new kit – and then seek to find enough new buyers to justify the effort.

Under Gilroy's tenure in the 1980s, the Range Rover was eased more upmarket and became more profitable. To support this, Land Rover marketing shifted and became more relaxed about promoting the Range Rover as a luxury limousine, a status symbol for people with an important job to carry out, be they royalty, the armed forces, diplomats, celebrities, the emergency services or the police.

As part of this strategy the interior was gradually revised. Plusher trim materials were specified, including velvet seat fabrics, Connolly leather, plush pile carpet and American walnut wood inserts, while factory-fit air-conditioning was also made available.

With the availability of automatic transmission and four doors, Gilroy's 'Project Eagle' could be brought into play. This would homologate the Range Rover to be sold in the North American market, something which had been ignored throughout the 1970s as supplies of the vehicle were so limited and BL's dealer network was in terminal disarray, with collapsing sales of the MG and Triumph sports cars. A completely new sales network was established, Land Rover North America (LRNA), and a series of improvements in build quality and interior trim were instigated to allow the vehicle to be pitched as a premium off-roader for the US market.

Mick Jones credits the establishment of this US sales network with a shift in attitudes towards design: "LRNA were a great support in the mid-1980s, with Design organising large presentations in the US. Design development was enhanced and departmental status started to be elevated as a result."

To support its 1987 launch in the US market, Evans set about

---

3. This led to a plan for Carbodies to utilise the Range Rover four-door body as the basis for the new CR6 taxi it was planning. In an advantageous deal, Carbodies agreed to tool up the shorter front door panel for free, in exchange for the rights to use it on its new model, but by 1983 this taxi project had been abandoned.

Independence and expansion of the range

To allow for air-conditioning, a revised lower instrument panel arrived in late 1983, featuring four rectangular air vents.

updating the Range Rover with a new plastic grille incorporating horizontal slats, new bumpers, and concealed door and bonnet hinges, to go with the more powerful 178bhp 3.9-litre V8 engine that was being developed.

Other interior upgrades over the preceding couple of years also constituted Eagle support activities, preparing the ground for the US launch. The biggest change was the revised lower IP that was designed for 1984 with four rectangular face vents and a larger instrument pack. 'Project Aquila' was a sub-part of this initiative, to install a new centre console in 1985 for the EFi version, followed by a redesigned two-spoke steering wheel from 1986.

In October 1992, a long-wheelbase LSE four-door finally made its appearance. As was often the case with the Range Rover, there had been numerous aftermarket LWB conversions, but the in-house project aimed to push the Range Rover as the king of luxury 4x4s, complete with a 200bhp 4.2-litre V8 and ABS brakes. There was also the addition of an entirely new air suspension system for the first time, known as ECAS (Electronically Controlled Air Suspension). This LSE model had a stretched chassis with a 108in wheelbase, with all the extra length going into the rear door, giving a more balanced look to the Range Rover, truly generous rear legroom, and far better access than the original four-door model.

Range Rover door cards underwent several redesigns as the model was pushed upmarket. Compare the 1983 In Vogue (left), with the plush treatment for the 1992 model.

Land Rover Design – 70 years of success

Comparison of the standard Vogue and the Vogue LSE. The 108in LSE was designed in 1990 by David Evans and introduced in October 1992. At the same time in 1992, the previous VM turbodiesel was replaced with the Discovery's new 2.5-litre direct injection turbodiesel as an option.

Independence and expansion of the range

The struggle for independence 1985-88

Under the 1983 reorganisation of BL, Land Rover had become part of Land Rover Leyland, a separate division from the Austin-Rover Cars Group, and included Leyland Truck and Bus and Freight Rover vans. The inclusion of Land Rover in this commercial vehicles group seemed surprising to many at the time but the thinking behind it was that it might allow the group to be sold off as a marketable venture more easily if they were separated from the debt-laden Cars Group. The pressure to privatise BL was a major policy under the Conservative government of Margaret Thatcher in the 1980s, with Jaguar being the first part of BL to be successfully spun off in September 1984.

The next stage in the privatisation plan was to sell Land Rover Leyland to General Motors. Talks began in spring 1985 to sell the truck, Land Rover and Sherpa van operations (but not the bus operation) to GM as a way of consolidating its mutual commercial vehicle operations in the UK – none of which were currently profitable. The proposal was for GM to take a 49 per cent stake in the company as a way of retaining some British control, which – not surprisingly – proved a sticking point for the US company. At the time, the government was embroiled in the Westland helicopter affair, and was concerned that a company with significant military contracts should remain under British ownership. After several months of talks GM eventually accepted this compromised deal, and a formal agreement was drawn up.

Two days later the Secretary of State for Trade and Industry, Paul Channon, informed the BL team that the deal with GM was not politically acceptable after all because of the risk to Land Rover's 'Britishness.' The deal was off. It is ironic that this point of 'Britishness' that seemed so vital to the UK government at the time has subsequently survived ownership by German, American and Indian companies without objection and has allowed JLR to flourish into the highly successful company we see today.

Next up was Ford. Discussions were opened in late 1985 for the complete sale of BL to Ford, which was particularly interested in the K-series engine being developed, and of course Land Rover, which did not overlap with any existing Ford of Europe products. Into 1986, BL was required to provide Ford with detailed financial information for these discussions as part of the due diligence process. Many executives on the BL side felt this was highly damaging to their situation, as Ford would gain access to future product plans and detailed plant profitability figures. Some of this commercially sensitive information was leaked to the press, and a political storm ensued with the Thatcher government. At this point the talks ceased, and Ford instead went on to buy a stake in both Mazda and Jaguar as a way of expanding its operations.

In May 1986, the government decided that a single chairman and CEO was required at BL to lead it into private ownership. The job fell to 52-year-old Graham Day, a Canadian who had helped with privatisation of British Shipbuilders and was well respected within government circles. Most of the previous board directors left the company at this point, including Harold Musgrove and David Andrews, and under Day the name of the company was changed to Rover Group plc.

Day decided that a period of calm was needed to restore confidence and that the diversion of management time to these privatisation matters had proved extremely damaging for the day-to-day running of the company. The position of Land Rover was revised so that it was more closely linked to the Rover Group, leaving trucks and Freight Rover as a separate division that could still be sold if a buyer could be found. Wisely, he kept Land Rover as a self-contained company on a single site at Lode Lane with its own product development team and design studio.

By the end of 1986, Day's new team had drawn up the outline for a deal with DAF to spin off a new company with DAF having 60 per cent of this new Leyland/DAF combination, which would relieve Rover Group of the whole truck and Freight Rover divisions. Into 1987 he also negotiated the privatisation of Unipart, Istel Ltd (Rover's computer services arm) and Leyland Bus, leaving a much more manageable Rover Group comprising Austin-Rover cars and Land Rover.

At this point – somewhat out of the blue – came an approach from British Aerospace to acquire a significant stake in Land Rover to boost their military operations but this was later widened to include the whole of Rover Group. For Graham Day it was the ideal scenario, a British solution to the privatisation agenda and one that had parallels with SAAB, Rolls-Royce or Daimler-Benz, where technology synergies across cars and aircraft had existed for decades. By December 1987 BAE made a formal approach to the government to buy Rover Group, at which point Ford also expressed an interest, which was quickly dismissed. Negotiations proceeded rapidly, and by March 1988 a deal was hammered out whereby BAE would pay £150m for the government's share in Rover, with the government writing off £800m of Rover Group debts. In return, BAE would agree to hold onto Rover Group for at least five years to ensure stability.

A new lifestyle vehicle: Project Jay

Following the ending of discussions with Ford and the appointment of Graham Day, the Rover Group Board was able to return to matters closer to home. In September 1986, there was a decision to be made over the next major investment, which was between an MG sports car project or a new Land Rover model spun off the Range Rover. There were insufficient resources to do both projects, and at that meeting the decision was made to go with Land Rover's 'Project Jay' over the MG.

Project Jay had kicked off around a year earlier as part of an initiative under Gilroy to look at advanced projects that could take Land Rover into new markets. A new team was formed called Swift Group, an in-house think tank to explore projects from feasibility to pre-production using reduced time scales. This was a six-man team housed in Block 46 at Lode Lane, under the leadership of Steve Schlemmer.

The group soon settled on the idea of a more lifestyle-oriented off-roader that would be priced below the Range Rover and offer a more modern vehicle than a Land Rover County Station Wagon. This would compete with established competitors for this type of vehicle, namely the Ford Bronco, Jeep Cherokee XJ, Nissan Terrano, Daihatsu Rocky, Mitsubishi Pajero and Isuzu MU.

The Jeep Cherokee XJ had been launched in 1983, and was notable for being a unibody design, rather than having a separate ladder chassis, as did all the other rivals. The Nissan Terrano had a choice of 2.4-litre petrol or 2.7-litre diesel engines and, like the Mitsubishi Pajero/Shogun and Daihatsu Rocky, was offered in three-door or five-door form. The Mitsubishi also offered a seven-seat layout in five-door form, plus the option of a smooth 3.0-litre V6 petrol engine as a top model (with a 2.0-litre turbocharged petrol engine or a 2.5-litre turbodiesel).

In Europe, GM was known to be developing another rival for Jay. This was the Vauxhall/Opel Frontera, a rebadged Isuzu MU 4x4 that would be assembled by Bedford in the UK. When finally launched in 1991, the Frontera

Land Rover Design – 70 years of success

was offered in three-door and five-door versions with 2.0- or 2.4-litre petrol engines plus a 2.3-litre diesel. It would be a major competitor, and Land Rover was keen not to be beaten to the market.

It was therefore becoming clear that a diesel-engined variant would be essential for Jay, and was likely to form the majority of sales, particularly in Europe and Asia. Indeed, it was this aspect rather than the body format issues, that dominated initial discussions within the company, and the diesel engine would go on to define Jay's position as a distinct model from Range Rover.

By early 1986 there was a clear agreement to base the new model on the Range Rover, and Design started preparing sketches and models to support the pre-programme planning that was required before the project could be offered up to Gilroy and the Board for approval as an official programme.

To keep costs under control the vehicle would use not just the Range Rover chassis and 4x4 drivetrain, but would also carry over the bulkhead, windscreen and doors. For designers Mike Sampson and Dick Bartlam this was a severe restriction, yet they felt the vehicle could regain its own character with the illusion of a shorter rear overhang, a vertical rear window and side opening rear door. George Thomson would be chief exterior designer for the project.

One key aim was trying to get away from the style of the Pajero and Trooper, and come up with a more characterful profile that could accommodate headroom for two side-facing rear seats in the trunk, not unlike those traditionally offered in the 88in Station Wagon. Eventually, a stepped roof profile was developed, with characteristic Alpine lights to give an airy ambience.

This Mike Sampson sketch from January 1986 shows one of the first ideas for Jay, with a Land Rover Station Wagon roof profile finished in a pale contrast colour. Alpine lights and lots of glass in the roof was an objective from the beginning.

Independence and expansion of the range

Further sketches by Mike Sampson.

Land Rover Design – 70 years of success

Definitive five-door sketch by Dick Bartlam.

A full-size clay model was started on a Range Rover rolling chassis in early 1986, and this was ready to present to the board in September. This clay model nearly had an unfortunate accident when being viewed outside by the designers just prior to the board viewing. "I am the first person to have ever driven a Discovery ..." recalls designer Alan Sheppard. "Being the smallest and youngest, I was asked to climb into the clay buck to steer it. As it got outside it tipped down the ramp. This thing weighed 2-3 tonnes with clay on it and it took off down the car park. Without brakes. I heaved with all my might on the steering as it was about to wipe out a line of cars. It went round in three large circles before it finally came to rest ...!"

In August 1986, the design teamed moved into much-improved studio facilities back at Lode Lane, where work on Jay and the new Range Rover continued. Soon after, Tony Poole decided to retire: the last of David Bache's team from the 1950s.

However, a new member of the team joined that autumn – Alan Mobberley. Another ex-Whitleyite, he had also worked at Jaguar with Thomson, and was brought in to lead the interior work on Jay, managing Alan Sheppard, 'Memo' Ozozturk and Mike Brogan.

The Jay project team was formally assembled towards the end of 1986. There was a core team of 50 staff, including designers, engineers and marketing staff. In charge of the project was Mike Donovan, a manager from Business and Product Planning, who had previously overseen into production the 'High Cap' Land Rover.

Independence and expansion of the range

First clay model in Drayton Road studio, spring 1986. The model was built as a rolling model on an actual Range Rover chassis. (Courtesy David Evans)

Later version of this same model, inside Drayton Rd. This side shows Mike Sampson's design proposal.

The same double-sided model was used for all exterior development. This is the model that careered into the car park with Alan Sheppard at the wheel!

The same model with Di-Noc, seen outside at Drayton Road. At this stage the constraint of retaining the Range Rover doors was not fixed.

Rear view of same model at a later stage. The unchanged Range Rover chassis and rear crossmember demanded a long rear bumper. Later design refinements cleverly sought to visually reduce this overhang.

Land Rover Design – 70 years of success

John Bragg was in charge of engineering, with Dick Elsy in charge of development.

A strict £25m budget limit was imposed on Gilroy by the BL board. With both the timing and budget being tight, a 'simultaneous engineering' workflow was employed for the first time in the company to bring Jay to production within 36 months rather than the normal five years. As a further expedient way of minimising tooling costs a number of proprietary parts were used from other BL models. These included rear lamps from the Austin Maestro van, headlamps from the Sherpa van, instrument pack from the Austin Metro, and Rover 800 air vents. These latter were chosen for their cylindrical form that was integrated into a distinctive transverse tube theme across the IP.

In contrast to previous Land Rover projects, Jay was a fast-moving project that reached sign off very quickly, with the exterior design frozen by February 1987. The interior design was started in late 1986 and, as well as the in-house models, two outside consultants were commissioned to produce proposals, namely IAD from Worthing and Conran Design in London.

David Evans explains: "We were looking at a new approach to interiors, and thought we'd look at IAD, and believed using Conran would be an interesting alternative as well. The initial presentation was mainly pictures of textures, of current themes within the design world – interior trends and mood boards. There were a couple of small schemes, including the requirement to package a high radio which was in the brief." This was something the IAD proposal had ignored so it was immediately out of the running."

> **Women and Design**
> Until the 1980s, design teams tended to be predominantly male, although Rover was unusual in having Pauline Crompton as a female colour and trim designer in the early 1960s. In 1987, Mick Jones recruited the first female designer for his colour and trim group, Karen Hind, followed by Kim Turner (now Kim Brisbourne) in 1988.
>
> With Jay coming on stream, Tony Gilroy was concerned the board might miss elements of feminine input. Rather than follow the previous pattern of inviting wives of prominent board members to review design proposals, he encouraged Design to recruit a 'Ladies Panel' to review and comment on colour and trim proposals prior to board reviews. In today's eyes this might seem horribly sexist, but at least it was a step towards acknowledging the importance of the opinions of the other 50 per cent of car drivers …
>
> "This panel consisted of female employees from within Design, but also working in 'close' areas such as marketing, materials laboratory and as director's PAs to maintain security," says Mick Jones.

Stripped Range Rover body used for interior development at 38A studio, 1987 (Courtesy Mike Sampson)

Independence and expansion of the range

The early Conran model was striking, but crudely constructed from foam and plastic extrusions. Note the roof console and rear view mirror. (Courtesy David Evans)

Later clay version of the Conran theme, reworked by Alan Mobberley's team (Courtesy David Evans)

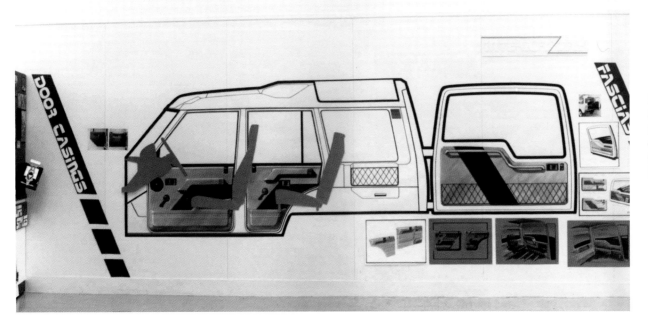
Interior tape drawing, done at 38A studio. Dick Bartlam redesigned the original Jay door casings theme from Conran Design. (Courtesy David Evans)

Land Rover Design – 70 years of success

Large-scale interior detail sketches focussed on new functionality. (Courtesy David Evans)

Independence and expansion of the range

Comprehensive colour and trim moodboard for Jay. John Stark loved using real items for moodboards: here from the worlds of fishing, yachting, and motorsport. (Courtesy David Evans)

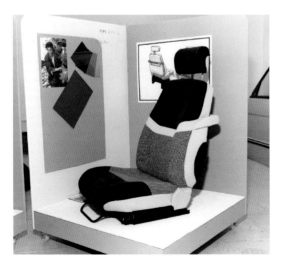

Jay seat trim mock-up. The idea to allow the seat trim to run over the bolster was novel and made the seat look wider. (Courtesy David Evans)

By March 1987, three interior models were ready to be presented: two in-house proposals and the Conran model. After some deliberation the latter was seen as the most interesting proposal to pursue, with its tubular theme and lozenge motifs that were fashionable at the time. However, as the project progressed, it became clear that Conran was unfamiliar with the complexities and constraints of an automotive interior, and needed substantial help from Mobberley's team to resolve the design for production. Mobberley confirms the Conran team had little understanding of the importance of draft angles for tooling, which created concerns in discussions with suppliers. "For example, the Conran model had raised pimples on the gearlever and interior grab handles that couldn't be moulded. We reversed it into a dimple that could," explains Mobberley.

In order to keep within the tight cost constraints, many plastic

Land Rover Design – 70 years of success

parts were low-cost vacuum formings, rather than injection mouldings. Despite this, Jay's airy cabin with a low belt line and tall roof allowed for a number of novel stowage solutions. Above the sun visors was a shelf for maps, and within the stepped roof area was a pair of stowage nets for children to use. The expense of the neat folding jump seats in the trunk was questioned by Tony Gilroy, but Marketing Director John Russell was confident they could sell at least 100 more vehicles with this seven-seat option. "Gilroy picked up on this number and stated that if the seat on-cost was more than the profit on 100 vehicles, then he could not have them. Fortunately for John the numbers worked in his favour, 95 being the figure, and Tony Gilroy allowed them," recalls Evans. The seven-seat option went on to be a huge success.

The low-cost tooling impression was offset by adopting a striking interior colour – Sonar Blue. "The Sonar Blue idea came from Conran, it looked very fresh," says Mobberley. An optional centre tunnel zip bag was also designed by Conran Design, which could be removed and used as a shoulder bag, and stowage bins in the trunk sides were a useful feature when no third row seats were fitted.

By September 1987 the interior was ready for sign off, and the project entered its final pre-production phase. The final budget for Jay came in at £45m; however, this included expanding production facilities at Lode Lane to build the vehicle, and the budget was still ridiculously modest even by 1980s standards.

The company considered several possible names for the vehicle, finally settling on 'Discovery.' Other names in the running were 'Highlander' and 'Prairie Rover,' before the decision to make Land Rover the brand, detaching it from any particular model. Launched on 16 Sept 1989 at the

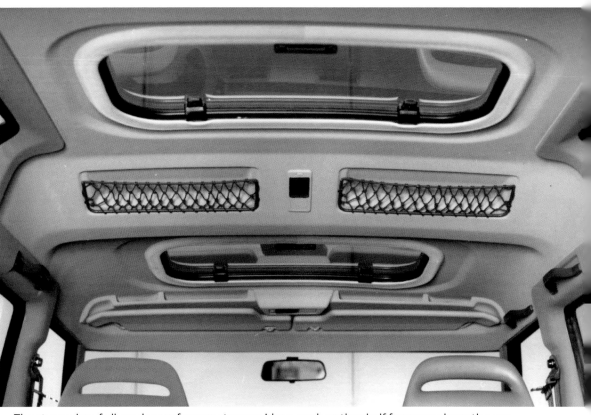

The stepped roof allowed room for new stowage ideas, such as the shelf for maps above the sun visors and a pair of stowage nets for children to use. Pop-out glass sunroofs were fashionable at the time and Discovery offered a pair of them, plus Alpine lights.

Production design of the Discovery interior in Sonar Blue. The Conran Design connection was promoted during the launch, although the actual design work had been executed by Alan Mobberley and the in-house team.

Independence and expansion of the range

Frankfurt Motor Show, it was priced from £15,750 for both Tdi or V8 petrol versions.

Product Planning were concerned that Jay should be deliberately positioned so that the overlap with Range Rover was minimised, and this helps explain the initial specification of the Discovery at launch. To keep the headline price down, everything not considered essential was an option. In addition, a range of over 50 fashionable accessories were developed, the first time Land Rover designers had attempted this in-house and far more comprehensive than anything else on the market.

Design-wise, there were several key elements that ensured it would not be seen as a cut-price Range Rover. The basic profile with a stepped roof, short rear overhang and vertical rear window was itself different, while the side-opening tailgate had the spare tyre mounted on it, rather than a two-piece split tailgate. The initial body was a three-door format, yet had provision for seven seats. Finally, there was the funky interior and promotion of the accessories range, not to mention a diesel engine as the main power unit.

Final three-door and five-door sign-off models in the viewing yard outside 38A studio, late 1987. Round headlamps were always the design intent, but the Sherpa square headlamps were substituted at a very late stage. (Courtesy David Evans)

Final GRP model, with decals. Note the sharper rear side window treatment from the production car. (Courtesy David Evans)

One of the original 'G WAC' press launch vehicles. Alloy rims were available only as a dealer accessory at launch, although decals down the flanks were readily offered.

Although designed alongside the three-door, the five-door Discovery was not introduced until 1991. That same year, the Discovery team was awarded the British Design Award.

Land Rover Design – 70 years of success

The worries that it would steal sales from Range Rover later proved unfounded, and customer surveys showed the two products were perceived very differently. After the launch of the car, the references to Conran's involvement on the interior were quietly reduced. As an early example of brand endorsement it had been a useful additional story, but was proving less beneficial after a couple of years in the marketplace.

Shortcomings in terms of materials and carry-over parts, however, were becoming more noted, and it was evident that a less 'lifestyle' approach to the interior might help drive sales. In response, a second interior colourway – Bahama Beige – was introduced in 1991 for the five-door, together with revised fabric patterns. By that time Discovery was getting into its stride, and production was ramped up from 300 to 500 per week.

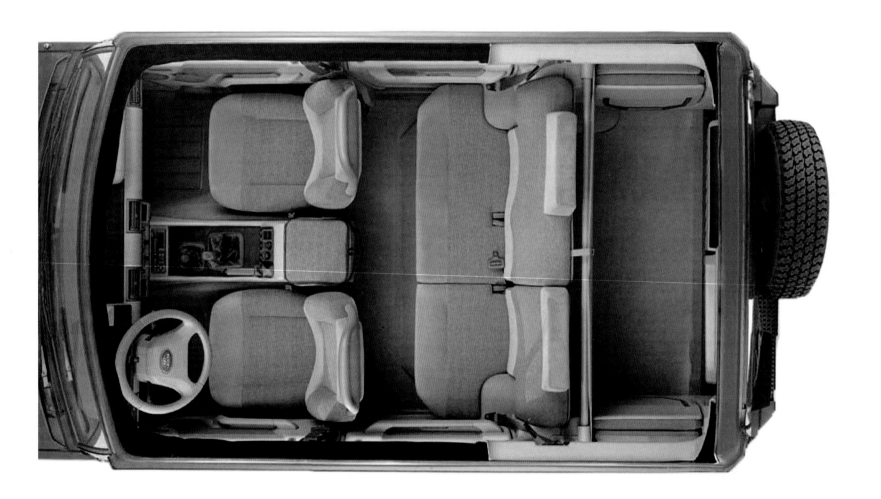

Discovery five-door gained a second interior colourway – Bahama Beige. The clever inwards-facing jump seats in the rear were designed by Mike Brogan.

1989-1996

Chapter 4

Canley studio and a new owner – BMW

Now that BAE was in control, product development at Rover Group was totally reorganised in 1988, with design and engineering operations being headed up by newcomer John Towers. By now Roy Axe had expanded his responsibilities at Canley, gaining advanced engineering and CAD teams reporting to him – a big turnaround from previous decades.

Up to that point the design function had been referred to as 'Styling' and the various studios were termed 'Styling Services within Austin-Rover.' Roy Axe explains the situation: "Traditionally in the UK, Engineering would frequently present us with a doubtful design with instructions to 'tart this up a bit,' which was a very annoying term to use. Styling was an appropriate title in those days in the sense that it was something that was applied after the design was completed. But the title changed to 'Design' when things got a lot more serious and our function began to have an input at the earliest stage of development."

Once Towers arrived to take over product development, Roy Axe's role began to be reduced. "The design department was going to take a lesser role, in fact returning to the old relationships at BL," recalled Axe. "This was going to put things back to how it was before I joined the company and I thought it was a completely retrograde step." The design departments would now come under an overall Product Development group, headed up by Dr Stan Manton.

As a result, Axe announced in summer 1989 that he would be leaving to set up a separate consultancy company based in Leamington Spa. Called Design Research Associates (DRA), it was initially tasked with developing aircraft interior design work for BAE, but Axe managed to negotiate a lucrative four-year contract to act as an outside studio for Rover car projects, with a pick of Canley staff, including Adrian Griffiths, Ian Beech, Jeremy Newman, Richard Carter and Chris Milburn.

Gordon Sked was promoted to lead the Canley studio, with responsibility not only for Austin-Rover but for Land Rover design as well. He had long been regarded as Axe's deputy, regularly shadowing him on his daily tours of the studios. Born in Kinross, Sked had joined Rootes in the mid-1960s as a 'pupil' rather than an apprentice, and had gravitated towards the styling studio, hoping to be given a chance as a stylist. A three-week probationary trial brought him into contact with Roy Axe, and convinced Axe that the young Scot possessed the necessary talent to enlist. Aged 23, he then joined Harris Mann in the Austin-Morris studio in 1970, later transferring to Rover in 1976, although he tended to work on Rover cars rather than Land Rover projects.

Under this reorganisation of product development, the Land Rover design team started to move over to the Canley studio complex, and the separate 38A Lode Lane studio was shut down. Most staff transferred to Canley during 1990, although those involved in the final stages of P38A project stayed to see it safely into production.

Integration of the different design groups was not without its struggles. There were several factions here. First, there were the

Land Rover Design – 70 years of success

Roy Axe

When Roy Axe was appointed Design Director in January 1982, he brought a much-needed professionalism to the product development process within BL.

Axe had two key advantages. Firstly, as a Design Director with international experience at Chrysler Corporation, he was familiar with the slick presentation techniques and fast turnover of projects that prevailed in Detroit. Secondly, as an outsider, he had little regard for the petty infighting across engineering departments and design studios that had been the bane of BL since the 1960s, plus he had been given authority by Harold Musgrove to change things as he saw fit.

Born in 1937 in Scunthorpe, Royden Axe had begun his career in car design as a teenager at Rootes in Coventry in 1953, aged 16. He was apprenticed through the rather antiquated techniques that were typical at Rootes, with wooden mock-ups rather than clay and limited sketch techniques compared with the Ford or Vauxhall studios in the UK. By the early 1960s Axe had proved himself as a designer and was fortunate to be promoted to manage part of the Arrow programme, particularly the 1967 Sunbeam Rapier derivative. With Chrysler taking control from the Rootes family that same year, Axe was again promoted at the young age of 29 to become Chief Stylist, where he set about building up a brand new design facility at the newly acquired Whitley site – the same building was later to become the Jaguar design studio in the 1980s.

Axe subsequently led the design team for the Hillman Avenger, Chrysler 180, Chrysler Alpine and Chrysler Sunbeam projects before moving to the US to join Chrysler's main Detroit studio in 1977. Here, he headed the interior design team, under fiery design chief Don DeLaRossa.

Although Axe was never given responsibility for Land Rover design, he remains a pivotal figure in the design story of Rover Group. Furthermore, he was a key figure in developing the world of design education in the UK, with far-reaching consequences that still resonate today. The Royal College of Art Vehicle Design course in London had been set up in 1967 under the sponsorship of Joe Oros from Ford UK but Axe was the second design chief to grasp the importance of the RCA course, regularly sponsoring two students a year from 1971. Early recruits who cut their teeth under Axe's RCA sponsorship included David Arbuckle, Geoff Matthews, David Evans, Gerry McGovern and Moray Callum, brother of Ian Callum.

Into the 1980s Axe was instrumental in developing links with the Coventry (Lanchester) Polytechnic Transportation Design course (now Coventry University). His support to develop the course and provide internships at Rover gave a number of undergraduates their first big break, including David Woodhouse, Oliver le Grice, Phil Simmons and Andy Wheel – all of whom went on to play further roles in the Land Rover story.

Talking to *Autocar* editor Steve Cropley in 2010, current Design Director Gerry McGovern said of Axe: "He changed my life. I wouldn't be doing what I'm doing today because he gave me a chance. I got to meet him and he sponsored me through college and got me to spend summers in his design studio at Chrysler. He taught me a lot about design. About form, sculpture and making a design work. Roy had a certain gravitas. He got the right balance of letting everyone know he was the boss but also knowing how to get the best out of people."

Former Design Director Geoff Upex agrees: "He turned Austin-Rover from a rag-tag and turned it into a professional organisation. I take my hat off to him. He projected the organisation as successful too. I joined simply because I felt Austin-Rover was going somewhere. He'd done a great job of putting design in the right place within the company. Roy was good at working through the vagaries of a pretty dysfunctional organisation and putting design in a significant position."

With his dark horn-rimmed spectacles and thick moustache, Axe could present a rather severe appearance. Like Bache before him, he had a distinctive dress sense that set him apart from his engineering peers, although it was heavily influenced by his love of American fashion with large check patterns and loud colours. During his Rover years Axe was able to indulge his love of Ferraris, first acquiring a 308, followed by two Berlinetta Boxers, then a Testarossa.

Axe became a full director of Rover Group from 1988 – the first time this had occurred and a tribute to his standing within the company under Graham Day. Following his time at Rover Group, Axe set up his own design consultancy, DRA, based in Leamington Spa. He eventually retired to Florida, where he died in October 2010.

Roy Axe.

Gordon Sked.

ex-Longbridge designers such as Adrian Griffiths, Richard Woolley, John Stark, Steve Harper and Michelle Wadhams still 'flying the A-flag.' Designer Steve Harper recalls: "At the time at Canley design studio there was only a few of us. There had been an invasion of 'Whitleyites' from the Chrysler studio when Roy Axe had taken over from David Bache, the likes of Gerry McGovern, David Saddington, David Arbuckle, and many clay modellers …"

Next, there was a bunch of talented young graduates, mainly recruited from Coventry Polytechnic, including David Woodhouse, Oliver le Grice, Phil Simmons and Jeremy Newman plus a group from the RCA including Howard Guy, Richard Carter and Chris Milburn. Other new managers had arrived too, such as Geoff Upex, who joined from Ogle Design in 1983, becoming Chief Designer of medium cars in 1986.

Finally there was the Land Rover design

Canley studio and a new owner – BMW

group, comprising another group of ex-Whitley designers, led by David Evans. "There were different camps," says David Saddington. "I knew Alan Mobberley and Dave Evans, I had seen George Thomson sketches at Whitley too. Old hands from 'the Austin' in Longbridge likewise remembered old Triumph colleagues, a few knew each other. But a bit of 'us and them' attitude. A few of the Land Rover guys didn't feel too comfortable since they were moving away from their close connections at the factory too."

Geoff Upex agrees: "It took a long time to integrate the Land Rover people into Canley, they felt psychologically different and it took a long, long time to get the crossover to happen." The difference was less to do with personalities and more to do with a difference of culture. The main Rover cars team had been working with Honda for over a decade by now, they had absorbed much of the Japanese philosophy and learned to cooperate on joint venture programmes with international engineers with an ethos of precision manufacturing.

By contrast, the Land Rover team had functioned throughout the 1980s as a totally British operation, with little influence from outside. To the Rover car designers, the build tolerances and wide shut gaps of products like the Discovery were a painful reminder of how they had moved on since the early 1980s. To the Land Rover team the wider shut gaps and large panel edge radii were born from functional necessity: they allowed the body to flex without panels chafing, and were a direct result of the tooling employed for the aluminium body panels. They also felt the wider gaps endowed the vehicles with a 'tough honesty' that fitted the image of ruggedness, something that was clearly inappropriate on, for example, a Rover 800 saloon.

At the same time, they had to develop projects such as the Discovery on pitifully small budgets, with many carry-over parts. Mick Jones comments that the Land Rover studio were hampered by financial constraints that only now became apparent to the team, which hardly helped internal relations: "It was surprising to learn that the Rover 800 door assemblies had cost as much as all of the Jay programme ..." Geoff Upex also gives them credit. "I admire them for what they did and make a success of it, for it was pretty damn crude," he admits.

Under Sked, Canley continued to work on major car projects with Honda, including the HHR (Rover 400/Honda Civic) and Syncro (Rover 600/Honda Accord). In addition, Canley developed its own R3 Rover 200 to bridge the gap between the forthcoming Rover 400-series and the Metro[1]. Now, with the Land Rover design group joining, the management structure was revised so that teams became entirely project-based, with Rover and Land Rover designers working together for the first time and David Evans's position as Chief Designer was revised. With Jay into production and P38A pretty much signed-off, Evans decided it was time to move on.

---

1. In 1993 a speculative project for a new Mini was kicked off, led by Dave Saddington with David Woodhouse, Oliver le Grice and Phil Simmons on the team as young designers. It was the first meaningful look at a proper new Mini (as opposed to a Metro successor) since the 1970s, and resulted in several proposals, including the pair of 'Spiritual' models – later shown at Geneva Motor Show in 1997. While there was insufficient funding for this as a serious programme with BAE, the timing was fortuitous and the preliminary Mini work proved a major point of interest once BMW was on board in 1994.

Canley design studio

The Canley design studio was situated on Fletchamstead Highway, part of the A45 southern bypass through Coventry. Following Roy Axe's arrival in January 1982, the old Triumph styling studio in Building 50 underwent a crash programme of expansion throughout that year with renovated offices, new studios and a new display showroom with turntables.

To reduce costs, the new studios were built inside the old Triumph assembly plant just next door, but the buildings were completely renovated with the latest surface plates and digital model bridges. To lead the enlarged modelling teams, Axe brought in Len Smith from the Rover Solihull studio as modelling manager. A model viewing yard was incorporated at the rear of the complex to present models in natural light, and allow the viewer to stand back from them. By early 1983, the Longbridge studio was closed, and staff were able to transfer to the new facility in Coventry.

"While no architectural gem it was a good practical layout and proved very workable for many years," commented Axe[1]. Other designers had somewhat mixed memories of it. Geoff Upex recalls "There were no windows, you never knew what time of day it was or what the weather had been outside, it was really strange." Designer Oliver le Grice agrees. "It was a bizarre place. The front offices for the design executives had wood panelling, a sense of faded Art Deco grandeur. But then you walked through a corridor and it quickly became like a second-rate comprehensive school, with peeling walls and horrible radiators. Then you walked further and realised you were inside an old factory. Quite interesting, almost untouched since day it closed – a bit like the Marie Celeste."

"The studio was probably the most vibrant place in the company, lots of enthusiastic young guys. It was a large studio, each designer had a big drawing table and behind were the older engineers on draughting tables with ties and pencils in top pockets. There were two or three surface plates with modellers – they were a complete bunch of characters. You could be pilloried as a young guy, made to look stupid. And they could shred you, one by one, day by day ..."

"There was never-ending whistling too – you never get that today. Or even breaking into song! Of course, everyone smoked, you just dropped the ash on the floor and the caretaker guys would come through with a big brush. It seemed a bit like Eastern Europe. But a very creative place all the same."

Early Oden clay model in Canley studio. The entrance block was a fine Art Deco building, complete with oak-panelled walls. (Courtesy James Taylor)

1. Roy Axe autobiography – A Life in Style.

Land Rover Design – 70 years of success

Defender and Challenger

Once the Discovery had been launched under the Land Rover brand, the company were presented with the problem of what to call the original Land Rover 4x4. Although badged as Ninety and One Ten models, these alone did not have a strong resonance with the buying public and a new name was sought. 'Defender' had been used as a former project code and was felt to have a strong image and, after some negotiations with another car manufacturer who owned the rights to the name, Land Rover adopted it as the generic model name for the classic 4x4 from autumn 1990.

The name change was made to coincide with the introduction of the 200Tdi turbodiesel engine from the Discovery, with the sub-names remaining but now being badged '90' and '110,' to indicate the Regular or Long chassis format.

At the same time, the 127's name was changed to the 'Defender 130,' although the wheelbase remained the same and the new figure was simply a tidying-up exercise. More importantly, 130s were no longer built from 'cut-and-shut' 110 frames, but had a dedicated chassis built from scratch.

The company also looked again at how the Defender itself might be replaced. Not only was the basic design now over 40 years old, but the 4x4 market was changing, with lower-cost Japanese models now dominating. A programme was begun to design Defender II, a vehicle that would be lighter and simpler to manufacture, using the 100in chassis and powertrain of the Discovery as a starting point.

The project was known as 'Challenger' and designer Peter Crowley was assigned to the project, based at Canley. The plan was to offer a full range of Soft Top, Hard Top and Station Wagon models, plus military variants, using as many of the existing components from the Defender and Discovery ranges in a modular fashion, in much the same way as the original Land Rover. The interior was planned to use the basic armature of the new IP design for the Discovery and Range Rover Classic, which would have provided a suitably modern yet rugged interior for the vehicle. "Challenger would have used a simple vacuum-formed skin, whereas the Discovery would be a foamed slush moulding," explains Alan Mobberley. "The armature was common, and we were given parameters of how much foam could be added."

"Two running prototypes were built off site, both went for military trials," recalls Crowley. "But it was a difficult programme due to the restrictions placed on it by the donor vehicle and carry-over parts. The idea was keep all of the chassis, drivetrain, screen, wings and all of the pressed steel inner bodywork, and we could change the outer aluminium panels. Marketing were very concerned about inroads of the Toyota Hilux pick-up into traditional Defender markets. Discovery was seen as a good fit for a car-based interior they were looking at, obviously Defender had always struggled with its out-of-date interior. I remember going down with Mike Gould to Earls Court agricultural show and we were looking around. At that time a tractor cab already had air-conditioning, sprung seats, stereos – by far and away more comfortable than a Defender. So, it was a case of trying to find a way of delivering that experience at low cost."

Photoshop images of Challenger as 100in (top) and 114in Double Cab variant. (Courtesy David Evans)

Challenger military prototype built in 1991. (Courtesy Dunsfold Collection)

These models were followed up by a military pick-up prototype in 1991 with a 114in wheelbase and canvas tilt, plus a side hinged spare wheel carrier as on the Defender. On this model the door was revised with a removable top half and new outer door skin profile that was carried through into the new front fender. It also featured a fold-down screen to offer a low-silhouette vehicle for battlefield duty.

"The plan was to deliver the Hard Top first, so all the effort went into that with a view to 'cut and shut' to make a Soft Top. But the engineers and I were concerned that we needed to do the Soft Top first to ensure the rigidity of the body, and to build-in features in the body that would allow much better canvas sealing than the Defender body with its 1940s features. For instance, other manufacturers were adding a threaded cable through the tilt, clamping it at the rear to create an air-tight seal, quick to add or remove. To try to fit that later would be bad news.

"The military wanted a true 130in variant to carry four ground-to-air missiles on a launcher, but the chassis couldn't be made strong enough to withstand the duration work cycles that the military were looking for. The other thing was the Discovery screen had a lot of plan shape. To have a military folding screen would require a bulky pantograph hinge to lie it flat on the bonnet. To the military, the bonnet is a valuable piece of real estate, it can be used for storing a spare tyre or standing on as a vantage point." Thus, to use this screen made little sense.

As the project developed, the costs to produce the full modular range began to increase and BAE were wary of the investment commitments. As with Llama, the final nail was the MoD itself.

Land Rover Design – 70 years of success

"It was never a true Defender, more a halfway house, not what the military wanted. They were tending towards much larger vehicles by then – the Steyr Puch or Hummers. They can carry lots of armour and payloads. The small runaround vehicle had fallen out of military favour."

In its place, a smaller project was started by Special Vehicles that revived the Cariba idea of a soft top leisure 4x4, using the existing Defender Ninety body, to be offered in the UK. Launched as the SV90 in summer 1992, this limited edition of 90 vehicles kept the full height framed doors of the Ninety, but added an external roll cage that extended in front of the windscreen. The sporty look was enhanced with Discovery five-spoke alloy wheels, a front bull bar, with brush bars over the headlamps, twin spotlamps and side steps. The 200Tdi diesel engine was unchanged from standard, however, meaning performance was hardly exciting. "But it created a huge amount of interest and did a good job of telling the world there was still a market for that vehicle," says Crowley.

As a further spin-off from the SV90, a unique model for the US market was developed to compete with the Jeep Wrangler. This ditched the Tdi diesel for the 3.9-litre V8 petrol engine that was far more suitable for American taste. Sold as the NAS 90 model, military-type Land Rover doors were fitted with detachable door tops and side screens in place of the SV90's full frame doors. The rear section of the roll cage was deleted, and a range of Surrey and

The Defender 90 NAS was fitted with a safari roll cage, brush bars and extra lamps. Almost 3000 were sold in 1994 and 1995. This example was used by Land Rover as a development vehicle, and has some non-standard parts. It is now part of the Dunsfold Collection. (Courtesy Dunsfold Collection)

Bimini soft tops was offered that emphasised the fun aspect of the vehicle. Wide 265/75 x 16in BF Goodrich tyres on five-spoke alloy rims were fitted, and the Californian look was completed with the addition of roof-mounted spot lamps and a range of vivid colours, including bright yellow. The NAS 90 gave a useful boost to Defender sales in the US and Canada from 1993 to 1997, when the Defender was withdrawn from sale in the US market, due to impending airbag and side impact legislation that would be impossible to meet without major modifications.

In parallel with Challenger, the company looked at ways of reducing weight using aluminium, leading to a series of lightweight concept vehicles. LCV1 was based on a Discovery with an aluminium chassis and body panels. LCV2 took a Defender but introduced a lightweight bonded and riveted spaceframe, while LCV 2/3 in 1997 added a substantially updated interior and aluminium body, with curved screen and side glass to improve aerodynamics. The project was not pursued further, but one example still exists in Gaydon museum

With the Challenger programme scrapped, new effort was put into upgrading the Defender to keep it competitive and enhance its appeal to more private customers, particularly in Europe. Fresher paint colours were offered from 1993, while glass sunroofs, tinted glass and cloth seats were increasingly available on Hard Tops and Station Wagons. In addition, Rostyle wheels and a range of new 16in alloy rims were made available across the range.

From 1994, the 300Tdi turbodiesel engine from the Discovery was offered which produced 111bhp and offered far more refinement than the old 200Tdi engine, increasing the comfort in the cabin by a significant amount.

More fundamentally, in 1998 the Defender was fitted with the all-new 2.5-litre, five-cylinder turbodiesel engine, badged the Td5. At this point the 200- and 300Tdi could not meet upcoming Euro3 emissions regulations so the Td5 replaced the Tdi as the only available engine for the UK and European markets.

A new Range Rover: Pegasus and P38A

Replacing the Range Rover was the biggest Land Rover project of the early 1990s for the studio. Two of the main players on this project were George Thomson and Don Wyatt. Thomson had started as an apprentice at Jaguar in 1967 and had risen to become a leading member of the styling team at the Browns Lane studio, responsible for the exterior of XJ40. He had then joined Chrysler's Whitley studio until its closure, whereupon he had briefly returned to Browns Lane, but found the engineering-controlled styling department too limiting for his ambitions.

Don Wyatt was an American designer who had studied at Art Center in California, and had then worked for Toyota and Chrysler. In 1981 he came over with Chrysler to the Whitley studio and had then run the advanced studio over in Poissy, outside Paris. In 1987, he heard that Land Rover was requiring more staff and applied to join

Decals

Adhesive decals were a popular feature on cars in the 1970s and '80s and soon made their way onto Land Rovers. Not only were they an inexpensive way to denote a special model, they also livened up essentially flat, featureless surfaces such as those found on caravans and boats. But why were they so prevalent on a Land Rover?

The real push came from Marketing, who loved them as a way of adding some low-cost pizazz, and gave them something to talk about at a time when there was actually very little that was new on the vehicles. Design was tasked with coming up with the ideas. "We did major decals, it was a Marketing request, there was a missed opportunity with the One Ten upgrade," explains David Evans. "A bit gaudy for many Land Rover buyers, but they were highly desirable and uptake was high."

Decal designs for the Land Rover changed each year, becoming more elaborate as they developed. The first County Station Wagons in 1982 had a straight four-band contrast stripe in cream on the 88in, with a two-band stripe on the 109in. The first One Tens used a similar four-band stripe too, while the Ninety County had a bi-colour kinked decal with a strobe effect towards the rear. For the launch of the Defender a very bold decal was devised with a brash 'DEFENDER' script running at an angle on the front doors.

Discovery 1 featured decals strongly at the launch, and offered several designs over the next five years that covered large areas of the body. Many observers felt they were garish and detracted from the Discovery's essentially clean lines. After the Romulus face-lift in 1994 they were dropped, with decals now being restricted to discreet badges or logos purely for limited edition models.

Decals never featured strongly on the Range Rover, although it did not entirely escape. The 1981 Range Rover 'In Vogue' limited edition had a contrast blue and silver side stripe running through the door designed by 'Memo' Ozozturk, while the 1982 Automatic Special Edition had a broader quadruple pinstripe in a dark contrast colour. One of the last models to feature decals was the 1990 CSK Range Rover, which had a discreet red pinstripe adorning its black flanks.

Decals on 1989 Discovery.

Strobe decals on the Ninety County Station Wagon.

Land Rover Design – 70 years of success

his old colleagues from Whitley. "I joined in October 1987 as design concepts manager, initially working on P38A," says Wyatt. "It was just starting, and we spent a lot of time on initial sketching. Pete Ludford had worked out the packaging for the vehicle and John Hall was project leader." At this point the new Range Rover codename was – confusingly – 'Discovery' – this was a couple of years before the launch of Jay that would take that title as a model name. "All the initial sketches had 'Disco' on them in fact,'" confirms Wyatt.

Sketch work began in late 1987 in the new Lode Lane design studio, with a series of large tape drawings. By summer 1988, six quarter scale models had been produced, including proposals from Italian design houses Bertone, Pininfarina and Ital Design. Two full-size clay models were then developed in-house, with alternative sides to them. In addition, the Land Rover management had commissioned a full-size proposal from Bertone.

"It was the most closed project and the most open project. In my opinion Gilroy and co were extremely nervous about replacing Range Rover," continues Wyatt. "They were willing to look at something new, but deep down all they wanted to do was to tweak the bumpers and change the door mirrors and call it a new Range Rover. We came in with a rainbow of options, and they ended up choosing the closest to the current Range Rover ... They were very uneasy all the way through. To them, we were bunch of strange people, a law unto themselves who all knew each other. It must have been a shock to the Land Rover organization to turn over the design of their most successful product to a bunch of outsiders."

The final theme review was in November 1988. "The Bertone model was eliminated," explains David Evans. "Themes B & C of the in-house designs were chosen to be amalgamated into one final design. George Thomson adapted his design to incorporate the changes and style approval was soon achieved." At this point Thomson took over the project and became production design manager. Key elements to the design were

Six Range Rover quarter scale models in 38A studio, summer 1988. (Courtesy David Evans)

Early P38A clay 'mass' model in the viewing yard at Lode Lane, early 1988. The Range Rover was an actual vehicle, with clay-coloured Di-Noc applied to make the comparison more valid. (Courtesy David Evans)

the clamshell hood, a floating roof, two-piece tailgate and a strong grille with horizontal bars. As with the late change on Jay, rectangular headlamps were deemed to be a necessary nod to contemporary trends of the time, although others subsequently felt this diluted the character of the vehicle.

The vehicle was originally meant to use a carry-over chassis from the LSE but as development proceeded it became an all-new design that utilised the latest CAD techniques to produce a stiffer chassis with reduced turning circle, although it retained the 108in wheelbase of the LSE. The bodyshell construction was updated employing a steel inner bodyshell with alloy skins rather than the riveted panel-on-frame construction used before. With Land Rover's in-house resources fully stretched on work with Jay, much of the body engineering was contracted out to MGA Developments in Coventry.

Once the Discovery name was decided upon for Jay, the Range Rover codename changed to 'Pegasus.' However, after this codename became known to the press and was bandied around in publications such as *CAR* and *Autocar* it was deemed compromised, and 'Pegasus' was thus changed to 'P38A' (after the design studio building) in 1990 in order to confuse matters.

The interior had been progressing steadily, too, and by early 1989 the design theme had been approved for production. One significant improvement was an all-new IP using high quality slush mouldings as on the Rover 800 rather than the crude ABS mouldings of the original vehicle.

Alan Sheppard was the main designer for the interior project, in parallel with his early involvement on what would become Freelander. He admits this is the project where he really cut his teeth as a designer, dealing with engineers and suppliers. "A lot of new technology was added. Navigation systems were on the horizon, also integrated car phones. Introducing all those complex electronics, Land Rover decided to go for the most complex system in world. But there were lots of firsts in that car, especially for

This Bertone proposal was shown as one of three full-size models in November 1988. A lot of this theme later resurfaced in the Citroën XM, also by Bertone. (Courtesy David Evans)

Theme C had a stepped belt line and conventional sill treatment. (Courtesy David Evans)

Land Rover Design – 70 years of success

Nearly there. This later clay model was a mix of Theme B front end and Theme C body side section and rear. (Courtesy Mike Sampson)

CAR magazine leaked details of the Pegasus, Pathfinder and Oden projects, leading to changes in project codenames. (Courtesy CAR)

an off-roader. For instance, twin airbags. But those first generation airbags were enormous to package."

The electrical system was based on that of the Rover 800, but the security system had reliability issues, leading to flat batteries. Body creaks and air suspension were other areas that could give problems on early models.

"We had nearly finished the interior; then realised the air-con was probably not powerful enough for the volume of the interior, so the whole IP and console was revised again to accommodate a much larger air-con unit. The final result was not as pure as the original concept."

In March 1994, the original Range Rover was also given a new slush-moulded IP to accommodate twin airbags for the US market. This 'soft dash' model continued to be produced as the Range Rover Classic in a bid to offer a lower-priced alternative to the new P38A model. In 1996, the Range Rover Classic bowed out after total production of 317,615 units.

The P38A interior was more car-like and sophisticated than the original vehicle and offered twin airbags. Beige and grey colourways were offered, with the console and upper IP in black. Crushed velour upholstery was bang on trend for 1994.

The P38A was launched with uprated V8 petrol engines of 4.0 litres, 185bhp and 4.6 litres with 225bhp (shown here). In addition, a new 2.5-litre 134bhp turbodiesel was sourced from BMW.

Land Rover Design – 70 years of success

The P38A was launched in 1994 at Cliveden House hotel. It was produced at the rate of around 25,000 per year. There was never any CKD assembly, and all 167,041 examples were produced in Solihull. The last one was built in December 2001.

Publicity shot for the face-lifted P38A typified the country aristocrat image that Land Rover was portraying by the 1990s.

Pathfinder, Oden and Cyclone: the Freelander

The origins of the Freelander are somewhat convoluted, since it was a project that was probing a new market niche, and would require a substantial injection of funding. The limited funds under BAE ownership were directed towards volume car projects developed with Honda, such as the R8 Rover 200, or towards sure-fire winners such as the P38A Range Rover. Investment was accelerated after BMW got involved, but Freelander development was protracted, and the final launch was not until the Frankfurt Motor Show in 1997.

Initial studies for what became the Freelander project began around 1988. Rover Group was looking at how it might enter a budding market for a small type of crossover vehicle that offered the style and image of a 4x4, but at a lower price and with a more 'fun' aspect. The Suzuki Vitara and Daihatsu Sportrak were the two main players in this niche market, with no European companies

Early Pathfinder sketches explored a number of possible formats, including monobox and more advanced styles for the project. This sketch is by David Woodhouse. (Courtesy James Taylor)

One of the clay model proposals for Rover Oden (Courtesy James Taylor)

Land Rover Design – 70 years of success

GRP mock-ups of Pathfinder and Oden around 1992. Much of the body from the A-pillar rearwards was common, but Oden used a car-like low front with typical Rover grille, whereas the Pathfinder had a deliberately bluff front end and roof bars. (Courtesy James Taylor)

A second 'Cut-and-Shuttle' simulator was built in late 1993 (by RSP). This was painted dark blue, with ride height increased by 4in over the Honda Shuttle. Rear lights were carried over from the Shuttle. Trimmed out with funky graphics and trimmed in a pink neoprene material, it was named 'Cyclone.' "We also called it the 'Stimulator,' it excited Sales and Marketing into doing the CB40 three-door," says Alan Mobberley.

planning to offer anything in this segment at that stage. Don Wyatt takes up the story: "Land Rover had Jay, they had Llama. As Advanced Design Manager I went off to develop a range of vehicles. We did a whole series of A4 sheets, filled two walls of sketches."

One European vehicle that had pioneered this niche had just gone out of production – the Matra Rancho. "PSA had looked at several vehicles like that as replacements," confirms Wyatt, who had direct experience of the Rancho during his period with PSA. Wyatt was subsequently dispatched for several months off site at consultancy RDS in Sheldon, with one engineer to work out a scheme for such a vehicle.

Over at Canley, Richard Hamblin had been put in charge of advanced concepts, and was naturally intrigued to see the results of Wyatt's early studies. "There were discussions about what can you do to a lifestyle vehicle that could be a people carrier, too. So we moved to Canley car park in some temporary buildings to develop it further," continues Wyatt. "We laid it all out, did a full-size tape drawing and some rough interiors. We then went into Canley main studio with this, and after two or three months we had done the first full-size clay, which was presented with good results."

For Rover it seemed like a great opportunity to explore ideas and become a 'first mover' in this lifestyle segment. But how best to approach this untapped market? To cover the breadth of possibilities, two separate lines of thought were developed.

The first – called Pathfinder – was to provide a fourth model in the Land Rover line-up, something a bit 'softer' than a Discovery and with a lower entry price of around £15,000. The second version was a Rover-badged five-door tall station wagon with 2WD only, not unlike the Mitsubishi Space Wagon or Honda Civic Shuttle of the time. This was codenamed Oden.

Both would use a unibody with independent suspension front and rear rather than a separate chassis and live axles as favoured by the Japanese models. The Land Rover project was devised as having two derivatives: a tall five-door station wagon version similar to Oden, or a three-door model with a detachable canvas rear tilt, like the small Japanese 4x4s. Both would use Rover transverse engines with 4WD and running gear based closely on the 200/400 series. Richard Woolley was senior designer on the Oden and Pathfinder, and recalls that, as the project progressed, the two styles became more unified until the differences were confined to the front end forward of the A-pillar, with the Rover Oden having a much lower bonnet. "I worked on the early five-door theme, also the three door, but was then seconded onto other car programmes."

To help 'sell' the idea to the Rover board, a driveable prototype was assembled that would allow the executives to appreciate the capabilities and get a feel for the concept far better than any static displays or clay models. This prototype was built by Clive Jones and Pete Ludford, and used a red Honda Civic Shuttle 4x4 that was cannibalised with a restyled body to produce a 'proof of

> **Rover Special Products**
> In March 1990, Richard Hamblin was made a director of a new think-tank that combined the talents of the previous Swift Group at Land Rover under Steve Schlemmer with those of the advanced design studio at Canley. Known as Rover Special Products, the group of 40 staff was housed at Gaydon technology centre, with a remit to explore options for niche products and take them to the point of pre-production.
>
> One early RSP project was the reintroduction of the Mini Cooper, followed by the Range Rover CSK – a limited edition named after the initials of Spen King. Next up were the Mini and Metro cabriolets, the MG RV8, and Rover 200 Tourer. However, the biggest project was Phoenix, which considered ways to produce a new MG sports car on a very limited budget. Don Wyatt and Hamblin were both deeply involved in this venture that occupied the next couple of years and resulted in a variety of concepts for MG sports cars, eventually leading to the MG F.

In 1985, Spen King retired from BL Technology. His final project had been to provide the engineering support for the MG EX-E concept, designed by Gerry McGovern and shown at that year's Frankfurt Motor Show. As the man behind the Range Rover, he was honoured with the creation by RSP of the Range Rover 'CSK,' a special edition of the vehicle named using his initials. (Courtesy BMIHT)

concept' runabout. The use of the Honda running gear was purely to get a prototype up and running within a few weeks, and was never intended as a serious proposal for the actual architecture of the vehicle. "We built the prototype in an extension to 38A studio, and tested it out behind the paint shop on some rough ground. It worked well," recalls Wyatt. Wags quickly dubbed it the 'Cut-and-Shuttle!'

Land Rover Design – 70 years of success

Later CB40 sketches by Gerry McGovern. (Courtesy James Taylor)

In the meantime, the Oden/Pathfinder project was put on ice for around 18 months and nothing further happened.[2] The investment required for the project was going to be substantial, rather more than the kinds of projects that RSP were handling, and sentiment was favouring a new MG sports car. Whereas in 1986 the MG project lost out to Jay, at this point the business case for the MG F was considerably more favourable than this unproven lifestyle project.

However, work recommenced in 1993 when RSP work was winding down. During that year, both Oden and Pathfinder projects were revived in tandem again under Geoff Upex, but gradually the consensus leaned towards the Land Rover version since this would offer a more sophisticated off-road ability, plus it could be priced higher accordingly, with greater profit margins. For Land Rover engineers it would be a useful exercise to develop their expertise in monocoque bodyshells, new plastics technology and electronic 4x4 control systems, but it would be a challenge to their skills to incorporate all the off-road capabilities expected of a Land Rover. For the Land Rover design team, it was the first chance for the Rover car designers to have a go at a Land Rover project, injecting some new blood into the team.

Following a major viewing in December, the project became known as 'Cyclone,' and there was agreement that the project should be progressed but the exact nature of the timing and the size of investment meant the project was not immediately approved.

All that changed in a matter of weeks. After the German takeover, the BMW board wanted to see what Rover had in the development cupboard, and was delighted to find the Pathfinder project so complete and ready to productionise. It was exactly the novel kind of project it expected Rover might have in place, and was utterly different from anything the German firm was currently developing.

The project was swiftly given concept approval by the board and Dick Elsy was appointed Project Director in April 1994, with Steve Haywood as Chief Engineer. Regarding the project name, it followed the precedent set by the Pegasus project by being named after the building in which it was developed. "We didn't want to call it Cyclone because Cyclone had been the concept vehicle and we were designing something new," recalls Steve Haywood. "So we decide to call it after the place we were based, which was Building 40 at Canley. Putting that together, we came up with CB40."

Gerry McGovern was delegated to oversee the reworking of the design from Cyclone to CB40, following completion of his work on the MG F sports car. By summer 1994, the basic architecture of CB40 was established using a monocoque bodyshell with independent suspension front and rear, mounted on subframes. In terms of engines, the car would have to use existing Rover units, which meant a transverse mounting rather than north-south as on all existing Land Rovers. A 1.8-litre version of the new K-series as used in the forthcoming MG F would be the main petrol engine, with the new 2.0-litre L-series direct injection turbodiesel as a second alternative. In addition, there were plans to install the 2.5-litre KV6, under development as a top petrol engine option targeted at the US market.

---

2. Allegedly, early prototypes of Pathfinder were shown to Honda in 1991 but the Japanese company declined to become directly involved in the project, preferring to develop their own project using the Civic platform. The resulting Honda CR-V was launched in 1995, some two years before Freelander but with styling cues that many observers felt owed a lot to the Land Rover. In truth, the whole lifestyle 4x4 market was well-established in Japan by the early 1990s with Suzuki, Daihatsu and Isuzu, and Honda were desperate not to miss out on this burgeoning market. More serious to Honda was Toyota's latest hit – the RAV-4 – which was the first compact crossover SUV, launched in May 1994.

Canley studio and a new owner – BMW

Full-size GRP model of CB40 three-door at Canley. The rear pillar was painted black at this stage, and the door drop glass shape would later be revised. Exposed fixings for the wheelarch mouldings was another idea being tried out. (Courtesy James Taylor)

GRP model of CB40 Station Wagon. A wave in the cantrail was added later. (Courtesy James Taylor)

Land Rover Design – 70 years of success

Interior sketch for CB40. (Courtesy James Taylor)

Under McGovern the design was refined, with more Land Rover 'toughness' being sought. Hence the rounded shoulder was enlarged and a more muscular front end developed, with an emphasis on blocky vertical elements either side of the grille, high-set headlamps and a pronounced skid plate under the bumper.

Young designers Alan Sheppard and David Woodhouse worked with McGovern on the exterior, while Sheppard also contributed ideas for the interior. Assisted by 'Memo' Ozozturk, the team promoted the idea of having very different characters between three-door and five-door interiors. "My interior was selected, the idea was about paring back, making it as minimal as possible. Making sure the structure of the car was visible from inside, lots of painted metal and exposed areas," says Sheppard. "I learnt most of what I know today on P38A, it was so intense, but it became useful on CB40 where we were looking for innovative ways to do things at low cost. For instance, the door cards were a minimal area and we used non-handed door bins."

The five-door Station Wagon was aimed at family buyers and would feature a conventional 60/40 split-folding rear seat and an interior choice of two fabrics, called 'Tapestry' and 'Canvas.' Three-door versions had a more youthful and fun feel to the interior, with two individual rear seats, unique door panels and seats trimmed in a jazzy 'Jungle' or 'Trek' printed fabric. The interior colour was a warm grey known as Smokestone, with an additional colour of Teal blue offered on the three-door only. In some ways, this strategy echoed the one used on the original Discovery, with the blue Conran interior used for the three-door derivative.

To add to its appeal, the three-door also came in two quite distinct styles. The first was a complete PVC soft top that rolled up to the roof bar for stowage to allow a totally open rear, like a Defender. This was known as the Softback. An alternative Hardback version was fitted with a detachable rigid hard top in place of the folding top. Both versions featured a pair of removable Targa roof panels, like those fitted to the Rover 200 Coupé, recently designed by McGovern.

Design development continued throughout the summer of 1994 to meet a tight deadline, to have the exterior and interior ready for sign off in November that year. By late 1994 the first engineering 'mules' were also built, using the Austin Maestro van body as a quick way of getting some initial testing carried out without having to wait for expensive prototype body tooling. The Maestro van was chosen as the wheelbase approximated that of CB40 and so allowed the driveline and suspension to be built into the vehicle to allow powertrain, chassis and suspension development testing to be carried out. 22 of these 'Mad Max' simulators were built and could be used on public roads without much need for elaborate disguise.

Roger Crathorne was demonstrations manager at this stage, and was one of many within Land Rover who had serious doubts about CB40's abilities as an off-roader, and whether it could be regarded as a true Solihull product. However, after trials at Eastnor Castle with Dick Elsy in one of the Maestro van prototypes, he was won over with its capabilities around the course, not least due to the combination of light weight, electronic traction control and hill descent control (HDC) features.

Six semi-engineered prototypes (SEPs) were constructed in spring 1995, with a further batch of 56 'D02' phase vehicles built in September. These were still hand-built vehicles off low volume prototype tooling and were used for the main development testing that was conducted throughout 1995-96 around the world in the harsh environments from Oman to Alaska.

Next up were D1 prototypes, which were built from production tooling in Solihull. Over 127 of these were built for crash testing and later development such as electrical tests. The final stage was a run of 'QP' and 'QC' pre-production vehicles to prove out the manufacturing processes and for quality confirmation prior to launch. Many of these were used for pre-launch press photography and dealer training purposes, as well as final tooling sign-off.

Ocean liners, peaches and coconuts: BMW takes control

The news on 31 January 1994 that BMW had bought Rover Group from BAE came out of the blue and many employees still clearly remember that moment. "It was a shock to us and Honda," recalls Upex. "I'd done three projects with Honda, and I was surprised by the kind reaction from fellow Honda designers. However, after BAE it was a dream. BMW were totally hands-on from day one."

BMW Chairman Berndt Pischetsrieder was the architect behind this takeover. Distantly related to Alec Issigonis, Pischetsrieder was a great Anglophile who believed the two companies would complement each other well. Rover would give a wider market coverage for BMW and bring some historic brands that were deemed under-exploited, such as Mini, MG and Land Rover. Dr Wolfgang Reitzle was head of BMW Engineering and had been involved with negotiations to supply the 2.5-litre turbodiesel for the P38A, so already had some knowledge of Land Rover's senior team.

At an early meeting of senior management Pischetsrieder

Canley studio and a new owner – BMW

Freelander Softback and Station Wagon publicity shot, October 1997. Initial production Station Wagons had black paint around the door frames, later deleted.

Freelander Softback. The five kinks in the sill survived from the earlier sketches and made it to production. The interior was more overtly funky, too, with 'Jungle' or 'Trek' printed fabric and Teal Blue plastic details.

Land Rover Design – 70 years of success

> **BMW buys Rover**
>
> By the early 1990s BAE was anxious to sell Rover and return to the core aerospace business. The financial position of BAE was proving to be much weaker than thought at the takeover, and in summer 1993 it approached Honda to propose that the Japanese company might like to buy a significant stake in Rover. Honda agreed to raise its stake to 47.5 per cent, but was not interested in taking majority control since it fervently believed that the company's best long term future was to remain in British hands, just as Honda would expect a struggling Japanese company might best remain in Japanese hands.
>
> By early January 1994, Honda believed it was close to signing an agreement on these terms and so was shocked to discover that BMW had made a counter-offer to buy the whole of Rover Group for £800m and that BAE would only accept an immediate cash offer from Honda for the equivalent amount. For Honda this was too much, not only in literal cash terms, but also in terms of disloyal trust after 15 years of close collaboration with Rover. Others have suggested that Honda was never interested in buying Rover, and that it knew too much about the inherent weaknesses in product development capability and manufacturing quality to want any deeper responsibility in the company.
>
> To Rover employees it was a shock, too. Although it was well-known that BAE wanted to sell Rover Group to a partner, there was the assumption that it would probably end up with Honda stepping in and taking a majority stake at the very least. By 31 January 1994 negotiations had been completed, and BMW took over the entire Rover Group.
>
> On the design side there was some celebration at the outcome. After all, the previous five years had been focused to consolidate the design identity and product line-up along the lines of a premium brand – the so-called 'Roverisation' of the Group, whereby the outdated images of Austin, Morris and Triumph were banished in favour of a single brand, much in the vein of BMW. To have that very same German brand come knocking seemed sweet justice indeed.

allegedly held up two different bottles of expensive vintage wines and proclaimed it was unthinkable that the two wines might be mixed together. Similarly, he explained, BMW and Rover were two strong and distinctive brands that should be treated with the same respect – that was the aim.

As part of an early exercise to explain the cultural differences between Germany and Britain, Geoff Upex put in place a presentation which looked at the contrast between British and German design. This was nothing new to Design of course who had previously examined this type of cultural difference as a regular part of design strategy with the Roverisation plan, trying to define the subtle nature of Britishness and how to communicate it through design. They had also used it as a tool to improve working relations with Honda colleagues on the joint venture projects, examining differences in Japanese and British design culture.

The design presentation showed interiors of traditional luxury ocean-going liners. The German vessels were thoroughly engineered throughout, with fittings that were functional yet high quality. The British ships focused more on visual appeal, with expensive fittings such as wood, leather and silk being employed in the passenger areas and emphasis on elegant lines, with the underlying specification of materials and the performance of engines being deemed secondary to passenger luxury.

A cultural training programme was introduced and one of the metaphors used was that of a peach and a coconut to demonstrate the differences between the two nationalities that might cause misunderstandings and friction amongst teams. The British liked to regard themselves as peaches: soft and pleasant on the outside but with a hard inner core that could be interpreted as less caring, possibly even stubborn. The Germans by contrast were perceived as having a hard, steely outer shell but in fact possessed a soft inner nature once you got to know them.

To begin with BMW allowed the current executive team of John Towers and Nick Stephenson to run the Rover side with little direct intervention from Munich. Meetings were all conducted in English and there was felt to be no need to introduce any German language lessons for designers and engineers: after all they all had the same approach to product development didn't they?

This proved deeply flawed. For instance, a BMW engineering manager would expect a more junior colleague to bring his skills to the task and simply carry out the allotted mission to the best of his ability, with the younger engineer fully understanding his place in the overall project. To the British, they saw this as arrogance: the German managers simply told them what to do and did not listen. They expected a more consultative approach whereby their views would be listened to and the senior staff would guide and mentor a project through every detail to an eventual conclusion. To the Germans this would be seen as insulting micro-management, implying that teams could not operate without their intimate involvement in the entire process, rather than setting targets and awaiting results from trusted engineers.

Under this new BMW regime the Land Rover designers and engineers gradually gained more respect. Previously their antiquated use of a separate chassis and body was seen in a poor light by their Rover car colleagues who had over a decade of Honda methodology drilled into them, complete with tight body tolerances and the latest lightweight materials. However, the Land Rover team still retained specialist body, chassis and engine development capability, areas that Rover Cars had allowed to decline, relying on substantial Honda support in most of these disciplines.

The Land Rover guys also understood very clearly the function and purpose of their 4x4 products and were able to show BMW how they had developed their own demanding engineering specifications to meet these specialist targets. This impressed BMW engineers and convinced them that they had something to learn from Land Rover, they needed to listen in order to design a 4x4 product that was fit for purpose.

As a result, there was a fear within BMW that the X5 E53 project might be cancelled and that all future 4x4 development would be handled by Land Rover. In the end this did not happen, both

Canley studio and a new owner – BMW

Gerry McGovern in 1995. (Courtesy BMIHT)

Land Rover Design – 70 years of success

Judge Dredd Land Rover

In 1994, Hollywood Pictures approached Land Rover to design and supply a vehicle for its latest film, starring Sylvester Stallone. Set in 2139 in 'Mega City One,' the film portrays a hostile urban environment. Society as we know it has collapsed, the traditional freedoms of society no longer apply, and the judges combine the function of judge, jury and executioner.

If you want to travel anywhere in Mega City One you hail a cab. These were liveried in yellow like a New York cab, but constructed like a fortress on wheels to protect you from the hostile world outside.

Designer David Woodhouse was dispatched to provide some sketches for the vehicle, despite being busy on CB40 work. His resulting design was highly menacing, with a streaked paint finish. The front had just one headlamp and a series of small lamps – like a one-eyed pirate. The body employed flat planes and mean windows, using influences from stealth bombers and American football helmets. Balloon tyres and wheel extensions completed the futuristic military look of the vehicle.

31 Forward Control 101in Land Rovers were converted for the film and fitted with GRP bodies using the Woodhouse design. Most were built up in the yellow City Cab colours, although there were also a number of 'City Utility' vehicles in silver, and 'Mobile Kitchens' in red and black.

Futura Design and Wood and Pickett coachbuilders produced the huge GRP body moulding, and Dunsfold Land Rovers won the contract to build the film fleet. Completing the order in the few months available and sourcing 101in donor vehicles was proving difficult, but luckily the British Army decided to dispose of a batch and these were snapped up at auction. Once filming was completed, the remaining Land Rovers were bought by Phil Bashall of Dunsfold Land Rovers, and many were converted back to 101in specification. Four were made road legal and used as promotional vehicles for the film, and a couple still exist.

Judge Dredd city cab.

sides realising there was room in this fast-expanding market for both the X5 and the Range Rover and that they would appeal to different customer groups.

The P38A was finally launched in September 1994. Despite the specialist Range Rover ethos, it was developed under tight budgets with BAE, and to BMW engineers it was seen as a less than ideal design, with many compromised areas that this full premium 4x4 should never suffer. "Reitzle couldn't get his head around it," recalls Upex. "He admired Range Rovers, they were his favourite vehicle. His view was this new car had been made down to a price, not up to a standard. It caused a lot of consternation."

Following Axe's move to set up as a consultant, Gordon Sked now decided to follow suit and establish his own independent design strategy company. "In March 1995, Gordon announced he was going away for six months, over to you," says Upex. "He spent the next few months in California, writing a report on the US market, which was delivered to BMW in September."

That year would prove a hectic year for Rover, with the MG F launched in March 1995, the 400 hatchback in summer and the new Rover 200 coming in October. With Geoff Upex now in charge, Dave Saddington was appointed Design Director for MG and Mini, with the 'New Mini' project (R50) getting official backing that summer. Heading up the CB40 development, Gerry McGovern was now made Design Director for Land Rover.

On the BMW side, Reitzle replaced Pischetsrieder as chairman of Rover and took a much more hands-on approach to the management of the UK operations. The following year Dr Walter Hasselkus took over from John Towers as CEO, the start of the process where German managers would replace British staff in many areas of the company. The one notable exception was Design, which remained free from this trend – a mark of its acknowledged competence within the group.

Canley studio and a new owner – BMW

Lode Lane production

During the 1990s, Solihull was reorganised yet again, with the Lode Lane site split between two main functions: Powertrain and Final Assembly. Powertrain produced V8 and Tdi engines in the North Works at the rate of 700 V8 per week, and 3000 Tdi per week. It also produced R380 gearboxes and LT230 transfer boxes, as well as complete powertrain units for Leyland DAF.

Discovery and Defender were both built in the South Works. The Discovery was built up from panels pressed at Motor Panels in Coventry. By 1994, Lode Lane was producing 1250 Discoverys per week – a far cry from the original level of 300 in 1990.

By contrast, the Defender was still built largely by manual methods, with very little mechanisation. The Defender body was assembled in parts, starting with the bulkhead being bolted to the chassis frame and other painted panels, then hung on the vehicle fore and aft. The modular construction of a Defender meant it was possible to have one line for all models and all variants. This was arranged in a C-shape within the South Works, with chassis and axles being assembled on the upwards leg and the body and final assembly occurring on the return leg.

With the P38A Range Rover coming on stream, the North Works was completely refurbished to facilitate final assembly of the new model. The BIW was produced in the East Works, with the body assembled from 260 panels supplied by Rover's Swindon plant, and built up at Lode Lane. The P38A marked the first time that robots were used at Lode Lane, for the main BIW frame jig.

A single paint shop was employed to serve all models, which was given a £3m upgrade in conjunction with P38A introduction to keep pace with the level of production, which had reached over 100,000 by 1995.

Finally, a £68m investment to allow for CB40 production was implemented from 1996. This saw a further eastward extension to the North Works to provide a new final assembly area for this high-volume model. By 2000 Lode Lane was producing 175,000 vehicles each year.

P38A Range Rover body assembly.

Land Rover Design – 70 years of success

Land Rover Special Vehicles (LRSV) department

From 1992, the specialist projects SVO department was retitled as Land Rover Special Vehicles (LRSV). This carried on the former commercial conversion work, but added a new limited-edition bespoke manufacturing service, such as building the Defender SV90. A limited run of 90 vehicles was produced from 1992-93 for the UK market. SV90s were painted in a new turquoise metallic, known as Caprice Blue, with a black canvas tilt.

LRSV also introduced an exclusive series of paint colours, leather trims and wood veneers for the Range Rover in 1993, under a programme known as 'Autobiography.' This followed the trend by other premium makers to set up sub-brands offering a bespoke service for customers who wanted something more exclusive than the standard colour and trim selection, and were prepared to pay a hefty premium.

BMW had set up the profitable 'Individual Line' programme, while Jaguar had set up 'Insignia' under its own SVO division. Autobiography was the start of the colour and trim team having the opportunity to showcase more individual finishes and new materials they had been developing and has grown to become a major part of the business today.

The Autobiography custom-building service was offered by LRSV from 1993 on the Range Rover only. Exclusive paint finishes, including painted wheels, were available – albeit at considerable extra cost.

Canley studio and a new owner – BMW

1996-2001

Chapter 5

Geoff Upex era and Gaydon studio

Development of CB40 continued throughout 1996 and into 1997. Under BAE, the plan had been to contract out the production of CB40 bodies to Valmet in Finland, much as Porsche did at that time with the Boxster, and a 50-50 partnership had been formed with the Finnish company. Completed shells would then be shipped to Solihull, where final assembly would take place – an arrangement not unlike that with Mayflower to produce the MG F bodies in Coventry. However, with BMW in charge, the plan was scrapped in favour of extending the North Works at Lode Lane, and accommodating CB40 assembly into that corner of the site. In addition, there would be a huge new paintshop facility to handle the vastly increased volumes.

CB40 was launched at the Frankfurt Motor Show in October 1997 as the Land Rover Freelander. When searching for a suitable name, Product Planning had considered 'Highlander,' but 'Freelander' was given approval just a few weeks before the debut. Honda had introduced the CR-V in 1995 in Japan, but it was only launched into Europe during 1997, meaning that Freelander was the only European model in this sector, something that would secure its place as market leader for the next five years.

In Europe, the compact crossover SUV market was still in its infancy, and very much the preserve of Japanese manufacturers with new models such as the Suzuki Vitara, Toyota RAV4 and Daihatsu Sportrak having moved the game forward. These models offered a wide array of optional equipment that shifted their lifestyle positioning as being more sophisticated than earlier 4x4 products, and this was another area that the design team needed to focus on.

The Discovery had been the team's first attempt to show the possibilities of offering a range of in-house designed accessories and equipment that dealers could sell. CB40 would take this to another level with a small team dedicated to designing a wide range and different sizes of alloy rims, nudge bars and protection grilles for lighting. At launch, CB40 offered more options and accessories than any other rival, including a fairly wild 'Body Styling Enhancement Pack' for the three-door version. With the Discovery, Land Rover dealers were now becoming accomplished at promoting and selling such optional equipment, and it was providing valuable profit opportunities for the company, as BMW had long-since found. Things had moved on a long way from the old approach of dressing up the vehicle with little more than a few decals and a pair of Lucas fog lamps …

Design moves to Gaydon

Under BMW ownership there was major investment in all areas of the business, something that had always been the bane of Rover Group and Land Rover since the 1970s. Up to 1997 BMW invested £450m in the new CB40 Freelander, £400m in the new R50 Mini, £400m in a new engine plant at Hams Hall in Birmingham, and £220m in new paint shops at Cowley and Solihull plants. In addition some £40m had been invested in staff training, an area that previous

Geoff Upex era and Gaydon studio

Geoff Upex

Geoff Upex was Design Director throughout most of the period under BMW and then Ford ownership. Born in 1952 in Redruth, his upbringing in Cornwall was slightly bohemian, which provided him with a rounded set of skills that informed his approach to design management.

"I was pretty much always into cars from a very early age. My father was into mechanical things, he used to race stock cars and speedboats, and so I used to go to all these races with him, so I guess I was destined to be involved in mechanical objects. My grandfather owned a couple of garages so my dad was bought up in the garage industry. Once it's in your blood it's difficult to get rid of. My dad was an incredibly practical man, he even built our house. He retired when he was 33 and started doing all sorts of jobs: a waterski instructor, then breeding chinchillas etc. His passion for cars inevitably transferred to me."

Initially aspiring to be an architect, Upex spotted an advert for a new course in Transportation Design at Coventry Polytechnic, where he studied from 1971, one of the very first students on that course. During his studies he did an internship at Ford in Dunton, although it was not in the Design studio but rather in Body Research doing roll-over tests on Cortinas. "They didn't allow us in the studio, students were deemed a security risk," he recalls.

On graduating, he then studied at the RCA in London, although not on the well-known vehicle design course but rather in product design, sharing a flat in London with contemporaries Richard Seymour and Dick Powell. From 1977-83 he worked for Ogle Design in Letchworth, a consultancy with a balance of product and transport design projects, including the Leyland T45 and Iveco trucks. His manager there was Richard Hamblin, and when Hamblin departed for Austin-Rover to head up interiors, Upex soon got the call to join him.

Initially appointed as Chief Designer for interiors, he became Chief Designer for medium cars in 1986, responsible for the R8 collaboration with Honda that was launched as the Rover 200/400 series in 1989. This was followed up with subsequent Honda projects such as Synchro (Rover 600) and HH-R (Rover 400), both of which required extended periods of working in Japan, together with designer Richard Woolley.

Following the departure of Gordon Sked in 1995, Upex was promoted as Design and Concept Director for Rover Cars, including Land Rover, Mini and MG. With the Ford buyout he became Design Director Land Rover, a position he held until his retirement in December 2006.

Geoff Upex.

The Freelander was launched with a wide range of accessories.

For the Freelander V6 versions in 2001, a slightly deeper front bumper with ribbed skid plate was introduced.

Land Rover Design – 70 years of success

managements had never paid much attention toward, particularly with regards to management training.

Research and Development had not been ignored either, of course. The previous arrangement of local engineering teams based in the build plants at Longbridge, Solihull and Cowley was steadily wound down, with much of it being located to Canley by the early 1990s. That site had now reached its limits in terms of space, and was in any case a makeshift facility that was nowhere near BMW standards in terms of layout or equipment.

In 1993, a new £8m British Motor Industry Heritage Centre opened on part of the Gaydon site, now renamed the British Motor Museum. This was the first time the public had any access to Gaydon, and opened up awareness that Rover had a major facility here in Warwickshire that was world class. The Heritage museum funding was a totally separate venture to that of BAE and Rover Group but it coincided with a plan for major development of the site and Rover started to use the museum facility for product presentations and media events.

£30m was invested in the new Gaydon Design and Engineering Centre (GDEC) in Warwickshire – one of the most comprehensive research and development facilities in Europe. The Design team moved to GDEC towards the end of 1996 and have remained there to this day.

Despite the investment, things were not running smoothly with BMW. The original plan to leave Rover under British management control had proved a major error, and both John Towers and Nick Stephenson were replaced with BMW executives to run the British operations. Walter Hasselkuss was brought in as CEO to take charge but was himself replaced by Walter Sämann in 1998, and by now the British management was increasingly marginalised. German language classes for employees were introduced, too.

Gaydon in early 1990s, before major expansion on the site. View looking south.

Gaydon Design and Engineering Centre (GDEC)

The 900-acre proving ground at Gaydon had been operating since 1979 and had been developed over the years with a high speed track, 36 miles of primitive roads and off-road test tracks. During the 1980s, a climatic wind tunnel, cold driveability chamber, ten engine test cells and semi-anechoic noise test chambers had been added, with a resultant staff of nearly 1000 engineers based there.

In 1992, the decision was taken to centralise all core design and engineering activities at Gaydon, bringing to an end the disparate grouping of staff at Canley, Cowley, Solihull and Longbridge. The site had many advantages. Gaydon offered generous space and flexibility for siting new buildings and, being company-owned, there were no land acquisition costs and site security was already in place. The new M40 offered excellent access from Rover Group's manufacturing plants based around Birmingham and Oxford.

The first phase of building started in August 1993, with the building of a new semi-anechoic chamber and other ancillary works. Weedon Partnership Architects and construction engineers SDC Group were contracted to draw up a masterplan for the site using simultaneous engineering techniques, mirroring those used in product development. This allowed the plan to be modified and adapted rapidly as plans developed, and also made sure the building had the flexibility to allow continuous improvement. Designers Oliver le Grice and Michelle Wadhams worked closely with Geoff Upex and architect Terry Lee to come up with a flexible, open-plan layout. The acronym FEAST was chosen to sum up the objectives: Friendly, Easy, Airy, Safe and Tidy.

"Our ideal was a studio where you'd never have to go outside to the viewing garden because there's so much natural light," explains le Grice. "I remember the old process at Canley to get a clay model off its pins, onto its wheels, onto trolley jacks – a ridiculous procedure to get it into this awful viewing garden. Also, how different the cars looked seen outside compared to inside, always a shock moment. The whole idea here was to have enough space and natural light that there would be no surprise when moving a model inside to outside."

Detailed planning started in June 1994, soon after the BMW takeover, with the first soil being cut in March 1995. Terry Lee designed the building to mirror the construction techniques of a Land Rover, with a steel superstructure clad in aluminium panels. A conscious decision was taken to exploit the functional elements of the building, such as roof support beams and ventilation ducts, as integral and visually interesting parts of the internal design.

The design block was laid out in a U-shape split into three areas: Pre-Concept, Concept and Development studios plus a colour and trim studio. Three flush-fitting 7-metre modelling plates were installed in the Pre-Concept and Concept studios and two 30-metre plates in Development, plus smaller plates rehoused from Canley for scale models and interior work. All were fitted with the latest Stiefelmeyer surface measuring bridges. To allow for unhindered access for external suppliers, the colour and trim studio was visually separated from the main studio area, with north-facing light and dark floors to simulate road reflections in the sills and lower bodyside of models.

The presentation showroom contained three turntables, front and rear projection screens and could be linked to the adjacent design conference room with accommodation for up to 200 people. The essential large viewing garden was sited next door to the north of the block, with a small 'Zen garden' in the centre of the 'U.' An adjacent New Model Development Centre housed the model workshops and 5-axis mill, all under the management eye of Len Smith. Model shops included a woodmill, trim shop and paint and GRP facility. In total, the design studio's working capacity was more than doubled over Canley.

A central street housing a café and trees acted as a public space within the high security building, providing informal meeting and exhibition areas. 650 engineers were housed in the 'Centres of Competence' areas on the upper two floors with linking bridges, meaning the various project teams could easily communicate. The total floor area of the studios, workshops and two engineering floors was 30,380sq m, with an additional 1660sq m new prototype bodyshell workshop.

The resulting Gaydon Design and Engineering Centre (GDEC) became home to over 1000 engineers and designers. The first design staff were on site in October 1996 and the £30m facility was officially opened by HRH The Princess Royal on 7 March 1997.

Architect's impression of the entrance foyer of GDEC.

The main foyer entrance to GDEC was extended and made more imposing at the insistence of Dr Wolfgang Reitzle.

Main street of GDEC. This also acts as an informal meeting area for staff and visitors.

Some fine sunsets can be seen from the huge west-facing windows of the main studio. Roof support beams and ventilation ducts were left exposed as functional elements of the building.

Land Rover Design – 70 years of success

View of GDEC and main studios from the test track. The two main studios are linked to form a U-shaped facility, with a viewing garden to the left, enclosed by high fencing.

Discovery II – Tempest

After the P38A Range Rover, Land Rover's next task was how to replace the Discovery – and here things were not so clear. A plan for an all-new Discovery project, known as 'Heartland,' was instigated with a focus on increasing presence in the important US market, but the programme would require several years to come to fruition, and something more short-term was needed.

In its place, the Discovery had undergone a mid-term face-lift for the 1995 model year, codenamed 'Romulus.' The front end was lightly revised with a new grille, smaller turn signal lamps and large rectangular bespoke headlamps, rather than the borrowed Sherpa van ones used before. The interior saw more fundamental changes, with a completely new IP style where the armature was shared with an updated IP for the Range Rover Classic. To meet latest safety expectations this incorporated a passenger airbag, together with a Rover 800-style airbag

1995 model year Discovery, codenamed 'Romulus,' introduced a revised front panel with larger headlamps and smaller orange turn signal lamps. The face-lift was managed by Mike Sampson.

Romulus gained a new instrument panel, designed by Alan Mobberley. The radio was mounted high up, away from the heat of the gearbox tunnel.

Land Rover Design – 70 years of success

Early sketches for the Tempest programme by Mike Sampson. The plan was to revise the three-door Discovery as a five-seater, and add a longer seven-seat version.

Tempest had the appearance of Jay, but drawn with a heavier pen. This is a first go at the full-size seven-seater, 22 September 1993. Seen here in Canley viewing garden. (Courtesy Mike Sampson)

Second stage clay model, March 1994. A chunkier body side section is developed, together with the first attempt at a higher bonnet to clear the new Td5 engine. (Courtesy Mike Sampson)

Three-door clay model with new body side section, Canley studio January 1994. The right-hand side shows the longer seven-seat version. 'Guts' and 'Supremacy' proclaim the banners in the background. (Courtesy Mike Sampson)

Geoff Upex era and Gaydon studio

steering wheel – the first off-roader to be sold with twin airbags. Switchgear was courtesy of the Rover 800 and Austin Montego, with large rectangular air vents replacing the former Rover 800-style barrel vents.

Much of the Romulus work was shared across the new P38A that was launched in the same year. Mechanical changes centred around the new R380 5-speed gearbox, which replaced the old LT77 transmission, and the 200Tdi turbodiesel engine was comprehensively refined to become the 300Tdi with 113bhp. The 3.5-litre petrol V8 was uprated to 3.9-litres, too, and in this more refined form the Discovery was offered for sale in the US market for the first time, giving dealers a two model line-up to sell.

After CB40 was given priority for funding under BMW, much of the engineering resources were centred on that project, and the full replacement for the Discovery was amended to become a comprehensive face-lift of the existing Romulus design rather than an all-new model. 'Heartland' had developed into a grand plan to replace Discovery with two vehicles: a flat-roof five-seater (L50) targeted towards the successful Jeep Grand Cherokee, while a seven-seater (L51) would take on the role for a larger stepped-roof Discovery replacement. In the end both these were scrapped, and a more diluted project was agreed, codenamed 'Tempest,' with a tight budget of just £60m.

One of the key aims of Tempest was to offer a better solution to

See-through GRP model for the Tempest in comparison with the existing Discovery, seen here in the Canley viewing garden. Note the differences from the production model: rubber seals for rear windows, and no black wheelarch protectors. Panel gaps are still large, although deemed tighter and more consistent than on the old Jay model. The three-door variant Discovery was axed at this point.

Land Rover Design – 70 years of success

The Tempest's exterior and interior were overseen by Alan Mobberley. This is the final sign off GRP model in the new viewing garden at Gaydon, autumn 1996. All skin panels were new except the tailgate pressing.

the seven-seat layout. Under new legislation the previous sideways-facing jump seats were outlawed in many markets, and extra length would be required to allow foot room for two proper forward-facing seats in the rear, complete with three-point seatbelts and headrests. The initial plan was to increase the wheelbase by several inches, but budget restrictions meant that the existing chassis would need to be carried over, with its 100in wheelbase that dated back to the original Range Rover. Hence, the extra room would have to be gained behind the rear axle, with a rear overhang increased by 125mm (5in).

Although that was never ideal for the proportions, the stance was helped somewhat by an increase in tracks of 50mm (2in) and the fitting of 235/70-16in tyres as standard on most models. The chassis was updated with an extra crossmember at the rear, and the rear suspension was revised with a Watts linkage rather than the older A-bracket layout. Active Cornering Enhancement (ACE) was offered on higher specification models using an arrangement of electronically controlled hydraulic actuators and anti-roll bars. This was combined with self-levelling air suspension on the rear

The Discovery 2 was announced in 1998.

axle only, which improved handling with high payloads and for towing. ETC and HDC from Freelander were added into the mix, too, giving the Discovery 2 an impressive off-road capability that endowed the model with a renewed credibility as the premium off-roader. The other big mechanical news was the introduction of the new 'Storm' five-cylinder 2.5-litre Td5 turbodiesel engine with 136bhp, also introduced on the Defender at this time.

Alan Mobberley was design manager for Tempest, with Mike Sampson assisting on exterior design. In terms of design, Tempest was a chunkier, more substantial interpretation of the previous Jay style. Door inner skins were carried over, but new outer skins with a heftier pull-type door handle were designed. The A-pillar was thicker, as were the creases down the bodyside. The increased track and wider wheels were covered by broad black wheelarch mouldings, while bumpers front and rear were deeper and more solid-looking than before. The grille had a thicker frame in body colour, and the front turn signal lamps and rear lamp clusters were all chunkier components.

Most obviously there was a longer rear fender, with the side window and Alpine lights bonded in flush rather than using an ugly rubber seal as before. The roof panel was longer with a cleaner line to the stepped roof and with stronger roof bars fitted. Finally, the tall Td5 engine demanded a new bonnet, although the difference from Jay is not immediately apparent.

With so much budget allocated to the exterior, the interior was only given a mild makeover. The IP introduced under 'Romulus' was retained, but the cabin was refreshed with new door casings, revised switchgear and a more plush style for the seats. Charismatic circular cupholders were also added to the sides of the centre console as part of the new focus on US market appeal.

Cutaway of the Discovery 2. Air suspension was introduced at the rear to allow for the increased payload with the longer rear overhang.

The Discovery 2 in its element at Eastnor Castle test circuit.

The Tempest interior retained the basic IP from the earlier Romulus programme, but added a passenger airbag insert, plus new seats and door casings. The greater cabin length and revised third row seats can be compared to the earlier Jay interior on page 94. Discovery interior colours and fabrics were revised, moving away from blue and beige to grey, black and beige.

2002 model year Discovery 2 with revised tail lamp graphics.

The Farmer's Friend

Land Rover might have found its way back into the heart of the farming community if the Farmer's Friend had been developed for production.

The project was an engineering-led exercise to develop a chassis with superb cross-axle articulation. Instead of accommodating all the articulation in the springs, the idea was to make the chassis frame as front and rear sections that would pivot around a central joint, through which the propshaft would pass.

The principle was not unique: such articulation is used in construction vehicles, dump trucks, and the huge Mercedes-Benz Unimog. But nothing like a small runabout that could be used on farms had been tried, something far more compact than a Defender or a pick-up truck. At the time, Gators and quad bikes were still fairly new on the scene and beginning to become popular, so the engineers focused on developing a small 'mule' based around that scale of farm vehicle.

The idea was developed by Land Rover Chassis Department that worked on new concepts. The team was led by Martin Dowson and Tony Spillane, based at Longbridge. A single prototype was built over six months in 1998 using parts from a disused Subaru Sumo 4x4 van, with a 1.2-litre three-cylinder petrol engine mounted under the load bed. The suspension, engine, gearbox and controls were adapted to fit into a chassis frame and the resulting vehicle was tested extensively around Gaydon during 1999, with astonishing results.

The clever part was the 'floating cab' concept, where the driver's seat and floor were mounted centrally on a sliding joint each side, high enough to allow the beam axles to articulate as much as possible, while remaining relatively level itself. In practice, a staggering one metre of wheel articulation was achieved, but the driver remained at the average of the front and rear articulations, meaning he remained on a reasonably even keel no matter what the wheels were doing, which gave it a sophistication over rough ground that was unique – not unlike the concept of the original Range Rover.

Later that year, the design department was given the vehicle to try out some ideas, and Paul Hanstock designed a few extras including the front panel, roll cage and spotlights that gave it an endearing insect-like character. The inherent rightness of the design stands out as a neat product that could have fitted well to the Land Rover brand. With the spread of quad bikes as a ubiquitous farming tool over the following two decades it is a pity that this concept was never developed further as an opportunity to offer a premium British product in this market.

Roger Crathorne arranged for it to be preserved at the Dunsfold Collection.

1998 'Farmers Friend' prototype. The driving seat was supported on the yellow tubular sliding joint, meaning the driver stayed relatively level. (Courtesy Dunsfold Collection)

L30 Range Rover

As mentioned in the previous chapter, BMW engineers were never wholly convinced about P38A as a flagship product, and set about looking at ways to improve it within months of taking over. One early plan was to install the BMW V12 engine in place of the V8 unit, meaning a longer nose would be required. Clay models were done to show the unhappy result – at which point it was abandoned.

Don Wyatt takes up the story: "Next up they decided to do a 1999 model year P38A as a limited face-lift, and I was asked to start a model. It was just bumpers and lamps – a very restrictive budget. At the same time, BMW themselves had started a P38A model with a lot of changes including a new interior, big changes. I went to Munich to look at it, followed by a later review with Reitzle at Canley for both models. Seeing both, Reitzle decided on the plane back to Munich he didn't want a face-lifted car – he wanted an all-new Range Rover."

At this stage a new series of BMW-style codenames was started with the Range Rover being project L30 – for Land Rover, 3rd generation Range Rover, 30 years on. Meanwhile, Rover car projects would be given R-codenames.

Wyatt was appointed as Chief Designer on this new project, reporting to Gerry McGovern as Land Rover Design Director. Work started in October 1996, and for the first year his small group occupied a temporary studio off-site at Motor Panels away from the main attention at Canley, but later one or two models were produced at the new Gaydon studio. Although he had been involved in the predecessor, Wyatt was unequivocal about its shortcomings: Talking to *CAR* magazine in 2001, he said "P38A was lacking in image and personality and we wanted to get the personality of the original back."

Up to this point, Land Rover design had seldom been subjected to outside competition and scrutiny in the way that other design teams have to contend with on a regular basis. As long as they came up with proposals that skinned the hard points of the package, they were deemed okay by the board, be it BL or BAE executives. Under BMW that cosy assumption went out the window, and McGovern's team had to up its game, particularly when it came to such a juicy project as the next Range Rover. After all, what designer wouldn't want to have that on his (or her) CV?

BMW had scores of highly talented designers who would love to have a crack at this project, and so design chief Chris Bangle insisted it was opened up to the whole design input of BMW Group. For the initial pre-programme stage in January 1997, scores of sketches were submitted, and around 15 scale models were produced. Six full-size models were made, and, of these, the proposals were whittled down to four exterior designs that were reviewed in August 1997.

First off was a model from BMW Designworks in California by British designer Marek Reichmann, with an edgy, functional design. The second model was produced by BMW in Munich, with a radical nose treatment and slim horizontal headlamps. The third model was by young Rover designer Phil Simmons, and drew on the Riva power

Land Rover Design – 70 years of success

boat for its inspiration, where the form has a suggestion of power up front and the tail tapers away to the rear. The fourth 40 per cent scale model was by seasoned Land Rover designer Mike Sampson, with a very graphic front end motif. All the models had similar Range Rover traits such as a wrap-around DLO, floating roof and castellated clamshell bonnet, while the British pair both featured thickly slatted grilles and round headlamps as signature elements.

The four exterior designs were scaled up to full-size clay models. At the August viewing in Gaydon, Simmons' model was chosen by the board, although Reitzle was sceptical about whether it really had the right ingredients. "Okay, you've picked this one [the Simmons model], but I'm not sure. I want to see it and the BMW model progressed further," he said.

L30 was a huge break for Simmons. "My full-size model was done in the new Gaydon studio on the front plate near the windows. I was very aware that this was my big opportunity to get a car into production, which was my ambition. I absolutely gave it my all. I vividly remember the day all the cars were out in the viewing garden. Dr Reitzle gesticulated very positively towards my model. I was watching through the window, 100 yards away, just hoping mine would be the one. To my delight it was."

A further eight weeks were spent refining the design, which proved a steep learning curve for the young designer. "One of the initial difficulties was that I imagined the girder-like bumpers could be separate from the body," recalls Simmons. "It turned out that would be obscenely expensive to deliver in production,

Phil Simmons' sketch inspired by the Riva speedboat. "It was about trying to capture all the power in the bows that flows to this elegant rear end. The boat tail effect," he says.

Geoff Upex era and Gaydon studio

so the eventual solution was an integrated bumper beam with a conventional plastic cover. It seemed tame to me at the time, I didn't want to let it go but after some soak time it started to look right. Now, when I look back at those separate volumes I think – well that was a bit naïve Phil ..."

The British team had won the design for the L30 Range Rover, but for the next stage of productionising, the entire process was to be done at BMW's Munich studio, since that was where the engineering would take place. Thus, the whole team were shipped out to Munich for the next two years. "I was going over weekly once we had selected the design," says Wyatt. "Chris [Bangle] took it straight into the studio, put it on a plate right next to E65 7-series clay, we started working right there in BMW's studio. The front of the plate was the exterior, the back of the plate was the interior clay. There was a roof buck too, everything happened together on that one plate."

"It was a fantastic learning opportunity," continues Simmons. "There had been the initial intent to share much with E53, but Land Rover's engineers explained to BMW senior management that the requirements of a true Range Rover were far beyond that which the stock X5 platform was capable of delivering. To their credit BMW understood and supported that view, so a lot of extra effort went into it. There are large chunks of commonality with the E53 in there but there were major changes to wheelbase, front wheel position and the height of the driving position relative to the ground – the command driving position. If you enable the command driving position then so many other things then fall into place. We got great support from BMW."

Full-size model of Simmons' design, selected by Dr Wolfgang Reitzle. This was one of four models shown at Gaydon, August 1997.

Land Rover Design – 70 years of success

Work on the interior also proceeded, under the direction of Alan Mobberley. Seven initial foam models were prepared for the August viewing by Reitzle in Gaydon. "They were done in tan and brown, his favourite colours," recalls Mobberley. "'It looks like a forest,' he said, 'all I can see is trees, I don't like any of them,' – and marched off. That gave us a huge problem." With just eight weeks until the next viewing, Mobberley swiftly selected two models done by Alan Sheppard and Gavin Hartley to develop as full clay models. These were then moulded as hard GRP models and trimmed. "We worked through the night for the next viewing. But at the last minute an alternative BMW studio model appeared that had been shipped over to be presented." Much to Mobberley's relief, "Reitzle came in, looked at the BMW model and said 'You can forget that one. These two are brilliant, which one should I choose?'" Hartley's proposal was selected and Sheppard was dispatched to Munich to see the project though to production.

Hartley's design was a charismatic theme of vertical stanchions on the sides of the centre console, running through to support the IP's 'canopy.' These added to the architecture of the IP, and, when treated as wooden elements, gave a structural authenticity to the use of wood that was utterly unique for a luxury interior. The end caps in wood echoed this idea, providing a functional support for the air vents and floor lights. "We were looking at it from a very nautical point of view," says Hartley. "We wanted to try to understand Range Rover customers and to try and surround ourselves with design images customers can relate to. We latched onto ocean-going yachts and the spaciousness that you would feel on board." Hartley even brought in a titanium yacht pulley as an example of how the interior should be conceived: functional, aesthetically pleasing and expensively engineered.

"There's a very different look in a Range Rover to what you'd expect in a luxury car," continues Wyatt. "We talked about Land Rover design cues and from the Series 1 Land Rover to the Discovery there are strong horizontal elements and vertical supporting elements." The design team also acknowledged the architectural approach promoted by David Bache that the IP should be essentially symmetrical, with the instrument binnacle let into the driver's side. It also referenced current architecture – such as that by Nick Grimshaw – as influences.

Designer Kim Brisbourne (formerly Kim Turner) took a new approach to interior colours and materials. "Range Rover is associated with wood and leather but we wanted to do it another way," she explains. While P38A used fairly conventional colours, Brisbourne proposed new finishes. Cherry wood veneer as used on expensive Mission hi-fi speaker cabinets, and alloy and parchment finishes were added to bring a contemporary British feel to the cabin. Use of a twisted pile carpet and a new 'foundry finish' texture on the centre stack added to the effect.

Instead of a separate chassis, the L30 bodyshell was an all-new design with a 'body-on-frame' structure that integrated the steel chassis frame into the steel floorpan. The bodyshell was a hybrid construction, with much of the upper body using aluminium pressings bonded into the steel structure. All round independent air suspension using MacPherson struts and double wishbones was employed, rather than solid axles – the

Key sketch of L30 interior by Gavin Hartley. The background references show teak, premium stacked Hi Fi decks and a titanium yacht pulley.

Continues on page 137

Geoff Upex era and Gaydon studio

L30 interior introduced new types of finish, including walnut and cherry wood veneers, alloy and parchment finishes and a new 'foundry finish' texture on the centre stack.

The IP top pad used a new grain, with a non-animal grain finish. This shows a 2006 example with later rotary Terrain Response controller.

Land Rover Design – 70 years of success

Skilful rendering of L30 by Mike Sampson.

The Green Book

Wolfgang Reitzle tasked Geoff Upex with producing a document to explain the design strategy for Land Rover brands. "I thought, well, that's an amusing comment, I bet there isn't such a book for BMW...! Anyway, we produced it, there are only 12 of these books in the world. It's a visual description of what makes a Land Rover a Land Rover, what separates a Defender from a Discovery from a Range Rover. But it also makes the point that this isn't a simple formula, you have to be a Land Rover designer to actually translate this into a real object and it's also subject to progression. In other words, this stuff isn't fixed, more an underlying set of ideas about the way we do our work."

"Although Wolfgang had asked for it I don't think we'd finished it by time he'd left BMW. When he arrived back with Nasser I took them through it, and they both understood it – L30 was a prime example of it put into practice. And it became important because the new Discovery was the first project we needed to do.

"It put us in a slightly awkward position. After all, Ford make the best-selling SUV – the Explorer. Why not base the Discovery on that? I thought, hell, we've been here before [with BMW E53]. How to persuade these guys that an Explorer is not a Land Rover? Part of the answer is in the book, when you go through it you realise Explorer does none of the stuff in here, particularly when you drive the cars and understand what a Discovery should be capable of doing. The Explorer driving position – the whole architecture of the vehicle – is wrong for Land Rover. Ford's worldwide engineering boss Richard Parry-Jones understood this totally, he completely understood why Discovery needed a unique approach.

"This document proved a very powerful tool to persuade them. We used it on almost a daily basis in meetings and I sent it to every Ford board member. Part of this was about trying to come up with a range of vehicles that had some cohesion, because the history of everything before then was of no strategy, simply a mish-mash of opportunities.

"Fortunately Ford top brass understood our aims with L30 as an exemplar. The most important bit was nobody tried to tell us how to do what we were doing. Unlike Jaguar. Jaguar had a problem in that nobody in Ford did not have an opinion on the brand and what should be done with it. So they left us alone."

A rotary Terrain Response controller was fitted after 2005 in the Range Rover and Discovery 3.

The Green Book – only 12 copies were produced.

first time this was adopted for the Range Rover, and allowing even greater off-road capability than P38A. To offer a really spacious cabin and good ingress for rear seat passengers, the wheelbase was increased to 2880mm (113.4in). With weight up 15 per cent, more powerful engines from BMW were specified: a 4.4-litre V8 petrol with 282bhp or a 2.9-litre turbodiesel in-line six with 174bhp, both mated to a five-speed automatic gearbox with permanent 4WD and a Torsen central differential.

To achieve the outstanding capability expected of a Range Rover, the engineering team had to set new demands over BMW's normal specifications for cars. Engines required extensive modifications to the oil system, seals and cam-drive pulleys to cope with extreme angles of operation and deep mud. New degrees of sealing for the multiplex electronic systems were required to cope with deep water, dust and grime, while the huge wheel travel and massive torque called for driveshafts of a spec previously unheard of.

The immensely strong body also required a 3.5-tonne towing capacity and towing eyes that could take a 'snatch recovery' – a brutal procedure involving a full 5.5-tonne yank to represent hauling the Range Rover out with a still-moving recovery vehicle. All this was very different to a BMW X5.

With the L30 project, the team rediscovered the abilities and associations that made the original Range Rover so unique and well loved. In that car, the luxury was not in the fittings, but rather in the core experience: the refinement, performance and outstanding go-anywhere capability. The command driving position not only imparted a feeling of security and dominance, but the low belt line made the driver look dominant from outside, unlike many other recent SUVs. The original Range Rover employed advanced technology to deliver that experience, too: coil springs, self-levelling suspension, permanent 4WD and a smooth V8 engine, and the new L30 replicated that distinction using the latest dynamic stability control and HDC to redistribute the torque and ensure traction was never lost.

BMW sells out to Ford

While BMW had bought Rover group, Ford had also been busy buying up brands. In 1987 Ford had bought a controlling stake in Aston Martin, followed by a full buy-out of Jaguar in November 1989. A decade later, Ford decided to buy Volvo, and by March 1999 had created Premier Automotive Group (PAG), amalgamating Jaguar, Aston Martin, Lincoln-Mercury and Volvo into a new luxury division.

Ford sensed that BMW was deeply unhappy with the whole Rover purchase, and started to cast around to see if it could once again acquire Land Rover. This coincided with turmoil at BMW when Pischetsrieder and a number of executives were forced out due to continued losses at Rover. The executives included Wolfgang Reitzle – who was immediately recruited by Ford to run PAG!

Of course, with Reitzle on board, Ford now had direct access and insider information on exactly how Land Rover was structured, including the profitability and future model plans. Reitzle would have been passionate about the forthcoming L30 programme, and no doubt talked Ford boss Jac Nasser into urgent talks with BMW to add Land Rover to the PAG fold.

Reitzle's influence ran deeper than that. He also recommended that Ford design chief J Mays should hire Gerry McGovern to become Design Director of Lincoln. McGovern duly departed, taking Marek Reichmann, David Woodhouse and Phil Simmons with him at the end of 1999. It would prove a short departure however: "Three months later Ford bought Land Rover – we were back in the family again," says Simmons.[1]

Indeed so. In March 2000 BMW announced it would dispose of Rover, with the Longbridge plant being bought by a consortium led by former Rover CEO John Towers. The Cowley plant and the R50 Mini project would be retained by BMW, while Land Rover and Solihull plant would become part of PAG, with Bob Dover as CEO. Central to the £1.85bn Ford deal was the entire Gaydon site, including the Heritage museum, plus delivery by BMW of the L30 project through to production. The deal was passed and became operational from 1 July 2000.

Geoff Upex recalls the moment in March vividly: "I had to fight to persuade Ford to take the whole design organisation as a complete entity rather than splitting people up. I think five people went to Rover and everyone else stayed, unlike every other part of the organisation." This caused problems over the next few months of due diligence as Ford could not understand why Design had not been split up. However, Upex managed to persuade Ford that it made complete sense, since it would have to start from scratch and Design would be the first part of the chain in that process, so it would need full staffing right now. A completely new strategy for Land Rover would be required regarding platforms and engines and they needed to hit the ground running from day one.

Richard Woolley agrees that the situation that week was desperately harsh. "Each and every one of us had the option of staying with Land Rover or BMW, or else with the independent MG Rover company. Within a matter of days all the Rover models and signage were cleared out and the Rover badges on the uniforms had to be scraped off. At the same time BMW removed all the models and mock-ups for the new Mini and took them back to Munich. Overnight it was transformed from a multi-brand studio doing R50 Mini, MG, and a whole series of Rover cars to a purely Land Rover facility. That was the beginning of the Ford era."

Upex continues: "The moment the Mini models went to Munich and Rover stuff to Longbridge I said 'You guys turn this into a Land Rover studio by tomorrow morning and start working on Land Rovers.' We converted the studio into a Land Rover branded studio overnight so when the Ford guys walked in they'd think 'Well this is

---

1. Reichmann was later appointed Chief Designer at Aston Martin, Woodhouse is now Chief Designer at Lincoln.

Land Rover Design – 70 years of success

The new Range Rover was launched at the London Design Museum in November 2001. After the Ford takeover, the L30 was renamed L322 in line with Ford's three-letter codename policy.

For the 2002 model year the Freelander was face-lifted with black bumpers and revised tail lamps. The Freelander tailgate always had a convenient electric drop glass.

the one place in the world where you'd design Land Rovers.' The main thing I was concerned about was that Ford shouldn't start trying to disperse the design input for Land Rover, as had certainly happened with Jaguar. I was determined that if something was a Land Rover it would be designed right here in this building. Nowhere else."

Upex admits they started working on a raft of spurious projects to fill the studio space, quickly cutting foam models based on old data from abandoned projects just to make sure it looked like a full Land Rover studio. "It worked. We held the department together unlike all other departments in the company."

A few days later Jac Nasser and Wolfgang Reitzle arrived. "I'm the one who greeted them," continues Upex. "Wolfgang puts his arm around me and says, 'Hi Geoff, long time no see.' I took them both around, we finally came to L30 – Wolfgang's baby. It was sitting on the turntable, and Jac was delighted with it. However, true to form, Wolfgang starts berating me about some small details he'd just spotted such as some visible rear suspension bolts and exhaust outlets ... Oh God, just like old times!"

Under Ford, the existing Land Rover plans were scrutinised and very quickly the studio was working on a whole new range of products. One of the earliest ideas was to take the Ford Escape to turn into a replacement Freelander. In the end that didn't happen, and the Ford EUCD platform was utilised for the replacement, called L359.

The L30 Range Rover development took fully six years with a colossal investment of £1bn under BMW, although this did include £200m in upgraded facilities at Solihull to increase production to 35,000 per year. Under Ford it was renamed L322, in line with Ford's three-letter codename policy and as a way of reasserting project ownership from BMW. Despite the change of ownership to Ford, Reitzle ended up launching his 'baby' at the London Design Museum on 22 Nov 2001, with the vehicle going on sale the following year.

For 2004, the Freelander interior was upgraded with new seats, a revised centre console, and door casings.

Chapter 6

The Ford years

At first, Ford kept Land Rover and Jaguar quite separate, but after a couple of years decided it needed to integrate the organisations, particularly on the product development side. The transfer of engineers between Whitley and Gaydon started to become a steady flow as programmes were assimilated, although the design studios remained quite separate under Geoff Upex and Ian Callum.

"BMW bought us thinking we had more engineering capability than we really had, we'd become quite dependent on Honda for engineering support," explains Richard Woolley. "They bought us not really understanding what the core capabilities were. They knew the design department was very capable, but when left to our own devices things didn't go as well as they should have. Whereas under Ford we became part of a huge business which was one slick unstoppable machine rolling on. We were pulled along with it and all the Ford processes came into play: product development, quality and manufacturing. We became indoctrinated quite quickly. These processes were already embedded within Jaguar of course, and, likewise, we knew a lot of the Jaguar designers. We were two teams, but with different missions."

In the years following the Ford buyout, the Gaydon studio really powered ahead with new projects. With the focus purely on one brand rather than the array of former brands, the design team could concentrate on deeper development of its vision for the future of Land Rover and Range Rover, and Ford investment was available to support a much wider line-up of models.

In April 2002, Dr Reitzle left PAG, and was replaced by Mark Fields. Fields took steps to amalgamate Jaguar and Land Rover more closely to reduce costs, creating Jaguar Land Rover (JLR) that year. Mounting losses, particularly for Jaguar, meant Bob Dover decided to quit in autumn 2003, though he stayed until April 2004 when Joe Greenwell took over.

Discovery 2 was given a mid-term face-lift in July 2002, codenamed L318. Design-wise, this comprised a new front grille and headlamps that aped the 'twin pocket' jewel style introduced on the new Range Rover the previous year. There was also a new front bumper design to improve the approach angle, and revised rear lamp clusters that now promoted the turn signals to the rear pillars. Other upgrades included revised suspension settings and a more powerful 4.6-litre V8 with 220bhp for the US market only. The whole programme constituted a £24m investment.

Discovery 3 L319
The replacement for the Discovery 2, codenamed L319, was a thorough reappraisal of the Discovery brand, and how it could be strengthened in terms of its image as a quite separate model from the new Range Rover. Geoff Upex takes up the story: "Doing the new Discovery was very interesting. L322 almost designed itself, we got to the end result very quickly. But the new Discovery was more complicated, trying to reinterpret that car in a more modern way was

Discovery 2 was face-lifted in 2002, with a Range Rover-style front. This was codenamed L318.

hard. The old car was a nightmare from a packaging point of view, with that 100in wheelbase." The design team therefore started from the packaging, and how best to accommodate seven people in a new family vehicle that made sense for serious off-road use, with decent departure angles and wading capabilities.

Designer Andy Wheel was a new member of the team, working under Dave Saddington. "We started sketching on 1 July 2000 – the day Ford took over. I'd joined in May 1999, spent a year on L50 and L51, the two vehicles planned to replace Discovery. There were real challenges in that programme. L50 and L51 were meant to be subservient to BMW's X series, then Ford overturned that assumption and we had a clean sheet of paper."

The new vehicle would share much of its running gear with the L322 Range Rover, including the generous 2885mm wheelbase (113.5in) that provided a really good basis for a seven-seat holdall vehicle. Unlike the Range Rover it still employed a separate chassis frame, but was updated with independent suspension all round, using wishbone front suspension rather than MacPherson struts as on the Range Rover. Base models still employed steel coil springs, but most versions used self-levelling air suspension front and rear, also as on the Range Rover.

This new 'Integrated Body Frame' construction was known as T5, with some parts shared with the Ford Explorer. Rather than a traditional ladder frame chassis and unstressed body as before, the L319 used a chassis where the deep side members curved outwards and were formed using the latest hydroforming process. The load-bearing body was then bolted to this chassis. If the adoption of a separate chassis seems odd after the successful unibody architecture of the L322, the reason is simple: T5 was planned to be used for a range of Land Rover products including a new Defender

Land Rover Design – 70 years of success

Initial sketches for Discovery 3 by Andy Wheel.

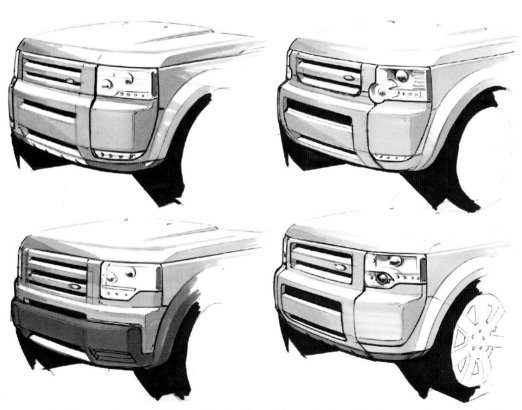

Sketches for front and rear end, focused on bold graphics and a functional approach to the detailing, such as headlamps and shutlines.

Andy Wheel led the design for the exterior of Discovery 3. He referenced design icons such as David Mellor cutlery and the Millennium Bridge in Gateshead as examples of timeless British contemporary design.

The Ford years

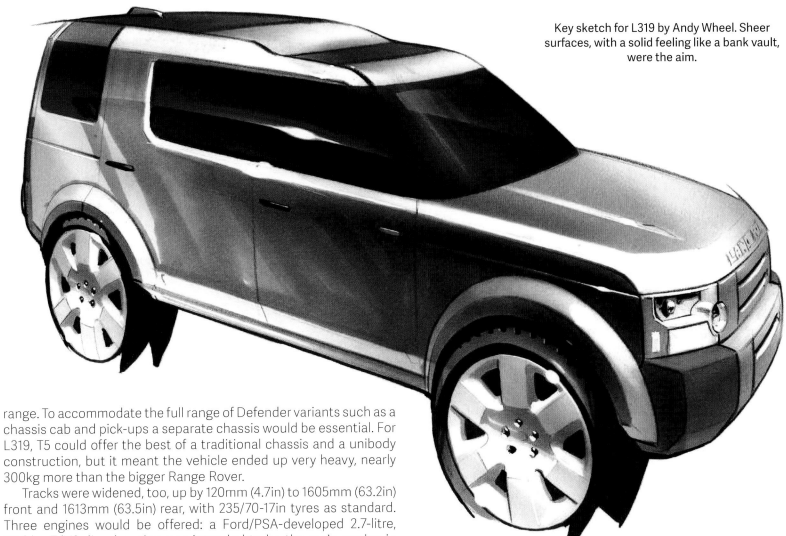

Key sketch for L319 by Andy Wheel. Sheer surfaces, with a solid feeling like a bank vault, were the aim.

range. To accommodate the full range of Defender variants such as a chassis cab and pick-ups a separate chassis would be essential. For L319, T5 could offer the best of a traditional chassis and a unibody construction, but it meant the vehicle ended up very heavy, nearly 300kg more than the bigger Range Rover.

Tracks were widened, too, up by 120mm (4.7in) to 1605mm (63.2in) front and 1613mm (63.5in) rear, with 235/70-17in tyres as standard. Three engines would be offered: a Ford/PSA-developed 2.7-litre, 195bhp TdV6 diesel engine was intended to be the main engine in Europe. For the US market and as the high-performance option elsewhere, a 4.4-litre petrol V8 of 300bhp was available, while an alternative 216bhp 4.0-litre V6 petrol engine was available in the US and Australia.

Certain design cues like the stepped roof and asymmetric rear tailgate graphic were kept, but the team still needed to decide how to get a Discovery that was modern and could sell across the globe. The outgoing L318 sold strongly in the UK, but struggled in other key markets, and the new model needed to appeal globally. "We wanted to put some controversy into the car too," continues Upex. "For instance, the implied line running through the door. We wanted this feeling of a bank vault passenger cell."

Andy Wheel produced a key sketch that won the day. "My pitch was for a 21st century Land Rover, wholly developed under Ford. I looked at modern architecture, something with a timeless feel. Appropriately detailed, but less is more. So much car design is fashion-led, but brand values can shine through by combining good proportions and a minimalist approach to produce a vehicle to stand the test of time rather than a 'me-too' vehicle. It carves its own niche."

No scale models were done for L319, Saddington's team preferring to try a new technique producing a couple of quick full-size speedfoam models using wooden armatures and polystyrene foam to mill out models that focused simply on volumes and proportions. "They looked big," says Wheel. "The new Range Rover was still a couple of years from launch – it also looked big. But we tried to optimise proportions, reduce the bulk to get a better balance between a spacious vehicle and not looking out of place on European roads."

Land Rover Design – 70 years of success

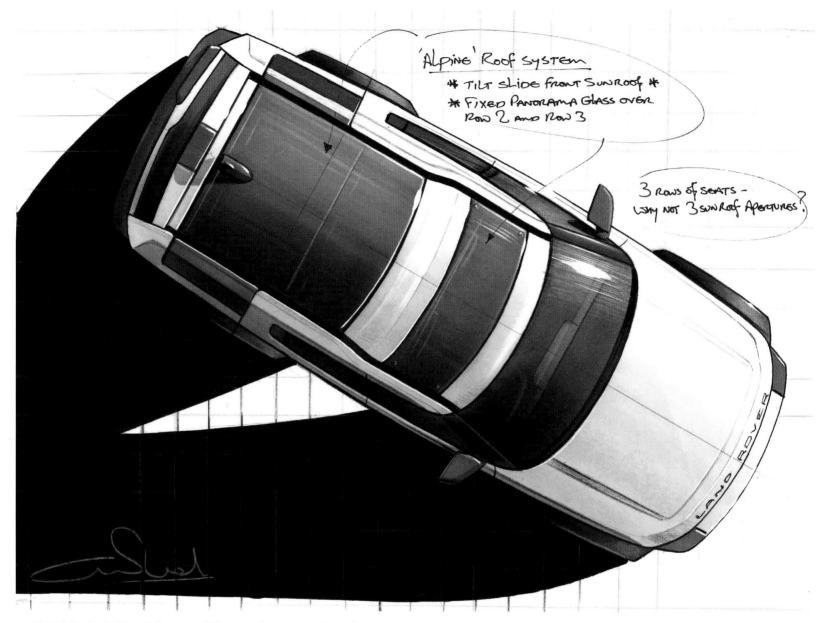

Sketch by Andy Wheel. Signature Discovery features such as the stepped roof, Alpine lights and glass sunroofs were a key part of the design theme.

The uninterrupted surfaces of Wheel's theme took a lot of work to get correct, since the entire bodyside is a single surface that needs to flow over the whole length of the vehicle. Any slight change in surface to accommodate a revised engineering or manufacturing requirement meant the whole bodyside had to be remodelled again.

"You'd get approached by some frustrated engineer working on the front fender because we'd changed the rear fuel filler bowl. Or we'd need to change the lower door surface because otherwise there would be too much paint damage by muck being thrown up the sides of the car," he explains. "Saying to the modellers 'We've got to do it again guys' was another thing. 'What? Again?!' they'd exclaim." French pop duo Daft Punk had a song called *One More Time* in the charts at that time in 2001. "So the modellers would be chanting that as they set off yet again. The biggest irony was Daft Punk's latest album was actually called *Discovery*!"

The rational, slightly minimalist design approach was not universally appreciated however. "Everything is there for a reason," explains Upex. "The 'form follows function' debate we used quite heavily. From our global research, people loved it or hated it, but that's what we wanted because sales volumes were not actually very

The Ford years

big, globally. We had a lot of strange reactions from women, that it was too brutal. Or, why is there only one fender air vent?"

The customer clinics were proving completely polarising. Wheel: "As a designer I like that, if you try to please everyone you'll end up with a mediocre vanilla design, whereas the clinic research was really different. It was all about the metrics, it needed to get a score of at least 5.8 out of 10 to make it past the next gateway. It didn't meet the score required." When the data for the exterior design was collated it was not a typical distribution bell curve, it was a U-shape with high scores at the extreme ends. "People either loved it or else said 'I don't get it, I absolutely hate it,'" says Wheel. "There was no middle ground there."

The design team came under a lot of pressure to change the design, to normalise it and continue the groove through the doors. Upex had to champion the design within the company by pointing out that the vehicle was still two years from launch and that tastes would change: new rivals such as the Volvo XC90 and BMW X5 were still fresh on the market.

"Marketing wanted a continuous line," explains Wheel. "By that time we'd gone so far on the surfacing that when we did actually put a continuous line then add a fillet on the clay model, the lines didn't look like they were joining up. So we deliberately moved the lines apart to make them look like there was an implied line. That meant that you couldn't then modify the doors, as there wasn't a

Discovery 3 at launch. The sheer surfaces with chamfered corners were successfully translated through to the production car.

Land Rover Design – 70 years of success

The asymmetric tailgate shutline was a neat demonstration of virtuous functional design. The right side allowed users to lean in closer, while the left side provided a deeper platform for sitting or for dogs to climb in.

continuous line there! To make it look right we had to disconnect the lines."

The interior followed the mantra of the exterior, taking a rational, functional approach. It built on the L322 theme with vertical stanchions either side of the centre console. It was a chunky architecture, a fairly masculine interior, although when trimmed with cherry wood insets it softened and looked more inviting. A robust 'dried river bed' texture was used on the IP top pad, a non-animal grain with lots of character to it.

L319 also featured a new 'Terrain Response' system. Previously, off-road driving had been a skill that many drivers found daunting. A wide-ranging knowledge of the vehicle was needed to be able to select the correct gear, transfer ratio, various differential systems and master various techniques required for tackling steep hills or deep water. Terrain Response attempted to take away as many of

The Ford years

Andy Wheel: "It was so rational as a piece of design that it provoked an emotional response when people looked at it."

Land Rover Design – 70 years of success

Discovery 3 in its element. The 'Integrated Body Frame' construction was known as T5, and used a chassis moulded by hydroforming. The load-bearing body was then bolted to this chassis.

The generous 2885mm wheelbase provided a truly spacious package for seven passengers.

the difficulties as possible. The driver selected one of six terrain types ('General,' 'Sand,' 'Grass,' 'Gravel & Snow,' 'Mud & Ruts' and 'Rock Crawl') on a rotary dial located at the rear of the console. The onboard computer systems then selected the correct gearbox settings, adjusted the suspension height, altered the differential lock settings, and even changed the throttle response of the engine suitable for the terrain.

The L319 was marketed as the Land Rover LR3 in the US Market. It was launched in April 2004 at the New York Motor Show, and went on to be built in Solihull at the rate of around 45,000 per year.

The Ford years

Charles Coldham produced the winning theme for the interior, which continued the rational form language of the exterior. The IP top pad used a new grain, dubbed 'dried river bed', a non-animal grain finish.

The Discovery 3 was heavily revised in summer 2009 for 2010 model year, when it became known as Discovery 4, or LR4. Most obviously, the front lamps and grille were revised, and rear lamps were redesigned with a pair of large white LED cylinders set within the red lamp housing. The big changes came with the interior and engines. Tight costings made under Ford meant not all the desired material finishes on L319 interior went ahead at launch. These were revisited in the 2010 revamp with significant improvements to perceived quality. The redesigned centre console contained simplified controls. The instrument cluster was updated with redesigned analogue speedometer and tachometer gauges for improved clarity. The analogue minor gauges of the LR3 model and the electronic information display were replaced by a single TFT screen capable of displaying information in a variety of modes and formats.

The engine line-up was revised with more powerful and refined engines. A new 3.0-litre 241bhp TDV6 Gen III diesel engine was offered, and a 5.0-litre V8 petrol engine with 385bhp replaced the 4.4-litre as the high-performance engine.

Discovery 4 had another refreshment for 2014: the headlamps gained the latest daytime running LED lamp signature and the grille and bumper were revised.

Summing up the Discovery, Upex says: "I still think it's an incredibly modern vehicle, unmistakable. Just imagine you're in North

Compare the 2010 face-lift interior. The chunky vertical centre stack was redesigned into a softer-looking centre console for the Discovery 4. Interior materials and colours were heavily improved to increase perceived quality. Door casings received a richer treatment, too. This photo shows a 2014 model with a rotary gearshift controller.

The revised Discovery 4 for 2010 model year introduced more body coloured parts. The body colour wheelarches had come in two years previously in many markets, but some regions – Russia, for instance – always preferred the black arches.

The 2010 changes transformed sales of the Discovery, justifying the name change to Discovery 4, or LR4. Optional parts such as side runners, body protectors, headlamp guards and roof rails were popular items for Discovery owners.

The Ford years

America, a market full of SUVs that all look pretty much the same. Why on earth would you go buy a Land Rover? There has to be a reason. Obviously one reason is that it doesn't look like a Ford Explorer or a GM Terrain. But you suffer from the problem that there are things that people hate, lines that don't run through, the asymmetric rear etc. But I'd do the exact same today if given a second chance." Andy Wheel agrees: "I'm still proud of L319, especially in the right colours and wheels. A really solid unpretentious vehicle, a very confident design."

Range Rover Sport L320

The Range Rover Sport was designed to give Land Rover a competitor for high-performance road-oriented SUVs like the Porsche Cayenne and Mercedes ML, but it retained all the traditional Land Rover off-road abilities. In many ways the key target vehicle was the BMW X5 Sports Activity Vehicle (SAV) which BMW had sought to develop in parallel with the L322, believing that a Range Rover should focus on supreme off-road ability as its key attribute. As such, it was ironic that Land Rover was now targeting that same sports-biased SUV niche now that they were under Ford control.

"Although it now seems an obvious model to slot into the range, at the time of its planning, there were many within the company who doubted the wisdom of producing it. It was a very difficult project to push through," says Upex.

This new model, codenamed L320, was based on the Discovery 3 T5 chassis but with 140mm (5.5in) chopped from the wheelbase. It was thus similar to the old P38A Range Rover in terms of wheelbase, length and height yet with much sportier proportions. This was combined with a more agile chassis with minimum body roll and lots of driver feedback for a sportier feel.

For the final 2014 face-lift Discovery 4, headlamps gained the latest DRL lamp signature, and the grille and bumper were revised. The new model naming policy saw the 'Discovery' name featured boldly across the bonnet for the first time. New wheels and body colours were also offered.

With the fitment of six-speed and eight-speed ZF automatic gearboxes, a Jaguar-style rotary gearshift controller was introduced on Discovery 4.

Land Rover Design – 70 years of success

L320 kept a conventional five-door layout. Despite the lower, more steeply-raked windscreen it still provided a roomy five-seater cabin.

Despite the wheelbase chop, under the skin it retained everything from the Discovery 3, including the double wishbone suspension used front and rear. According to Upex it shared 85 per cent of content with the Discovery by value, despite the two vehicles looking quite different. Because of that it proved to be hugely profitable for JLR.

"The Company was petrified it would cannibalise Range Rover sales," says Upex. "But we did a lot of research of the three cars together – Range Rover, Range Rover Sport and Discovery. It was obvious that Range Rover Sport and Range Rover appealed to completely different people. Women said I'd like one of these [Sport] but I'd love to know the man who has one of those [Range Rover].

That sums it up incredibly well!" In the end it did not cannibalise sales of Range Rover at all, nor Discovery. As the research indicated, it appealed to quite different customer groups, namely more women and more urban buyers.

The interior was designed Mark Butler and Wyn Thomas. "It was mainly done in CAD, only later did we do one in clay, probably one of the first IPs done that way. Foam models were milled out and trimmed directly for the first evaluations." L322 shared the basic IP armature of the L319, but added a new flowing centre console, rectangular air vents and unique door casings. In keeping with its sporty aims, the gearshifter was offset towards the driver.

The Ford years

Interior of L320 shared the basic architecture of the L319, but added a new flowing centre console design and rectangular air vents in the style of the Range Stormer concept.

As with the L319, the vehicle was substantially revised in summer 2009 for 2010 model year. The most significant changes were to the powertrains. Two new Jaguar-derived engines were offered, a 5.0-litre 375bhp V8 naturally aspirated engine, and the 5.0-litre supercharged V8 with a full 510bhp. The exterior was face-lifted with signature LED headlamps and a new two-bar grille, together with a new front bumper and fenders to create a more sporting and aerodynamic stance. At the rear, new light clusters and a revised bumper design echoed the smoother front-end.

The interior was comprehensively redesigned, using higher-quality materials, soft-touch finishes and superior craftsmanship to create a true premium look, befitting the Range Rover moniker. Whereas the Discovery had strong vertical elements overlaid on the horizontals for its theme, here the full-width horizontal beam of the Range Rover was allowed to overlap the console for a cleaner look.

Range Stormer concept

"It was a hell of a lot of work to get Range Rover Sport through the process. There was lots of opposition, which was one reason why we did the Range Stormer showcar," explains Geoff Upex. "Apart from the style, we were trying to change the emphasis of the brand away from off-road onto more on-road capability as well. We did it to show

Land Rover Design – 70 years of success

to the public that this brand isn't what you think it is, it actually has this other capability of on-road dynamics, too."

This was the first Land Rover concept ever shown at a motor show, and a great opportunity for Upex's team to show what it could achieve. Porsche had recently launched the Cayenne, and the idea of a more sporting SUV was still a novel concept in the luxury marketplace. Work on the show car started in May 2003 under the guidance of Richard Woolley and the project was done fast to meet the deadline of a launch for the Detroit Motor Show in January 2004.

The team did not spend a lot of time fiddling and tweaking the design for this Land Rover GT. "Doing it that quickly maintained its freshness, we didn't spend hours philosophising about details and concepts. We knew what we wanted and went out and achieved it in a very Land Rover way," continues Upex.

The exterior built on the styling theme for the L320 Sport, but exaggerated it to suit a shorter format with a pair of dramatic 'Blade and Runner' butterfly doors to access the four-seat cabin. The upper door swung up Lamborghini-style, while the lower door swung outwards to offer a convenient step up into the cabin, both powered by complex hydraulic systems.

Mark Butler and intern Ayline Koning worked on the interior. "We worked on the clay interior over the summer, it was very hot," recalls Butler. "A fantastic opportunity though, you always dream of working on show cars. I was doing the production surfacing for L320 Sport at the same time as Stormer."

"The first challenge was how to create a sports vehicle that hangs onto this command driving position because the two are diametrically opposed," he continues. "Sports cars tend to close you in, cocoon you from the outside world, whereas with the command driving position you get a very good, clear view of what's happening." The dilemma was solved by having a sporty, straight-arm driving position, a swooping centre tunnel and a high-set gearshifter offset towards the driver.

Designed in summer 2003, the Range Stormer featured dramatic 'Blade and Runner' butterfly doors, and sat on 22in alloy rims. Renderings based on 3D digital models were becoming the norm, using Alias and Maya software.

The Range Stormer show car was presented at the Detroit Motor Show in January 2005, finished in a striking candy-flake metallic paint colour with a hot flip called Oh!Range. "It had to be startling to get you noticed at Detroit," says Upex.

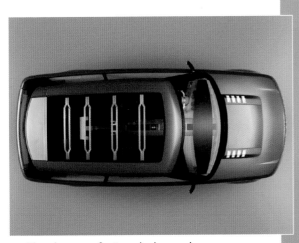

The show car featured a huge glass panorama roof supported on 4 aluminium spars. Built by Stola in Turin, the Range Stormer now resides in the British Motor Museum at Gaydon.

Land Rover Design – 70 years of success

The interior introduced the 'sports command' driving position, complete with a swooping centre tunnel and a high-set gearshifter.

In addition, the driver was seated in a deep hammock-like seat. This was designed by Koning to represent a Mobius Strip, with the cushion and seatback flowing together into the headrest in one continuous movement. The seat was constructed on an aluminium frame with four layers of thick saddle leather laminated together on a thin GRP armature to form the seat surfaces. The edges were left in a raw, unfinished state to give an appearance like Scandinavian plywood furniture. "We sourced the leather from a local saddlery in Kenilworth," says Butler.

The show car was built by Italian coachbuilder Stola in Turin. The vehicle was built on a shortened P38A chassis, with the clay models shipped out in August and work completed by December.

Upex sums up: "It wasn't about the production Range Rover Sport, it was about the Range Rover brand. In the end it had more coverage than any other car at Detroit. Unfortunately, it was later damaged by a moronic truck driver, who blew all the hydraulics when loading it. By then Stola had gone bust so it was never repaired."

Thus, the Range Stormer concept was developed as L320 was being finalised, and the production vehicle was launched exactly one year later at the Detroit Motor Show in January 2005.

Freelander grows up

With Range Rover Sport and Discovery 3 coming on-stream, the manufacturing space at Lode Lane plant was getting crowded. The decision was therefore taken in 2003 to shift production of Freelander to Jaguar's Halewood plant, and use up spare capacity there: the first time a Land Rover had been produced in the UK outside of Solihull.

The flowing console previewed the theme of the forthcoming Range Rover Sport. Light-coloured leather was used below the belt line.

The Ford years

Deep hammock-like seats were constructed with four layers of thick saddle leather with raw, unfinished edges.

PAG quarterly forums

Under the PAG setup, Ford Vice President of Design J Mays would come and visit the various studios individually every few weeks, to see how each project was progressing and to add some input. In addition, the design chiefs of the various PAG, Ford and Mazda marques would meet every quarter for a Design Directors' Forum to present their projects to Mays and to each other. These events were usually at PAG headquarters in London or at one of the other studios. This was in marked contrast to the lack of inter-marque communication that had existed under the British Leyland regime back in the 1970s.

"All the design directors and J would go through everything we were working on," explains Geoff Upex. "He would take input from everyone in the room. It was kind of tough but if you were doing a reasonable job you could get through it without too much difficulty. If not, then ...!"

Upex confirms it was a useful forum. "There was J, Ian and Moray Callum, Gerry, Peter Horbury, Martin Smith – some of the best designers in the world I'm sitting with and they're all giving you an input. There's not many places you could go where you'd get that input from people you respect, who really know what they're talking about."

Ian Callum agrees. "It was good fun, very amicable," he recalls. "Everyone was respectful of each other. It was a huge catalyst to us understanding each other's problems and understanding the high levels of design we had to reach."

Callum admits the forums were also useful to understand the values of each premium marque. "Now it's just Gerry and myself. I miss this interaction with other teams. There's nothing stronger than a fellow designer telling you that's not very good, because you do listen to them. I miss them actually – I miss those days."

Upex thinks the Gaydon team did well under Ford. "I think Ford have been a bit maligned, but they were fantastic for Land Rover. They left us alone, we just got on with it. We did what we thought was right, and they were pretty much happy with what we did."

This coincided with a face-lift for the Freelander, done under Richard Woolley, with new front bumper and lamps for 2004 model year. The signature element was the black stripe under the headlamp that gave it a 'tiger's eye' look. At the rear, the bumper became body coloured with supplementary lamps moved up high into the bumper corners to reduce the chances of them being obscured by mud. The cabin received an upgrade, too, with a revised top pad for the IP, a squarer instrument binnacle, new door panels and seats, while the centre console got the 'Range Rover look' with silver end cap stanchions.

By 2005, with the collapse of MG Rover, supplies of 1.8-litre K-series and V6 engines from MG Rover Powertrain had dried up and only the TD4 BMW engine was available for the last year or so of production.

As Land Rover had hoped, the CB40 Freelander was one of the pioneers of the compact lifestyle SUV market, and had proved an unqualified success with sales of around 75,000 per year since the launch in autumn 1997. As predicted, the market was growing strongly, and, as a European premium brand, the Freelander was still the model with the most authentic off-road credentials.

The two closest rivals to the Freelander Station Wagon were the Jeep Liberty/Cherokee, launched in 2001, and the BMW X3, which had arrived in 2003. Like CB40, the Jeep had permanent 4WD and was highly capable off-road, with a proper reduction transfer gearbox, whereas the BMW was far more road-biased than the baby Land Rover. In the US market the Chevy Equinox and Ford Escape/Maverick were similar transverse-engined off-roaders with part-time 4WD capability.

Outside the US, Asian brands still dominated the scene in many markets, with a host of new models launched in the early 2000s. The Mazda Tribute, Suzuki Grand Vitara, Subaru Forester, Nissan X-Trail, Hyundai Santa Fe and Kia Sorrento had all arrived on the scene, while the third generation Honda CR-V and Toyota RAV-4

Land Rover Design – 70 years of success

The face-lifted L314 Freelander was announced in August 2003, with the 'tiger's eye' front face. This is a three-door Sport.

were both due in 2006. In short, although it had done well over its eight-year life, the Freelander needed to sharpen up with new technology, materials and design if it was to remain competitive in this fierce new market.

Under BMW, the next-generation Freelander would likely have used the new X3 platform with north-south engines and RWD bias. With Ford now in control, the plan was revised. As the steel monocoque bodyshell had been all-new and was still relatively competitive, the initial plan in 2001 was to carry out a straightforward reskinning of the existing CB40, updating the engines and features as necessary.

However, at the end of 2002 this idea was scrapped in favour of utilizing the new Ford EUCD platform. EUCD was a beefed-up development of the compact Ford C1 platform used by the Ford Focus Mk II, C-Max, Volvo S40/V50 and Mazda3. In EUCD form it would be used for the mid-size CD345 Ford Mondeo, Ford S-Max, Galaxy, Volvo S60/V60 and S80 saloon. More pertinently, it would underpin the forthcoming Ford Kuga and Volvo XC60 compact SUV models, with transverse engines, six-speed gearboxes and 4WD capability using a new Haldex torque-sensing system.

This project, codenamed L359, continued the transverse engine layout of CB40, but provided new architecture that enabled a great deal of hidden component sharing among the Ford and PAG brands.

For Land Rover, this modified EUCD steel platform, known as LR-MS, would offer big savings in development. The basic floor structure,

The Ford years

firewalls, electrical components, brakes, and air con were shared, meaning a great deal of engineering development could be pooled and time spent on component testing and crash performance could be saved. Costs would be shared across the various Ford brands, with vehicle systems such as braking and electrical systems needing just one main supplier development programme. In addition, crash performance of the body could be commonised in certain areas, rather than designing unique solutions to front-end crash structure or airbag deployment, for instance.

Richard Woolley was in overall charge as chief platform designer, with Earl Beckles as lead exterior designer and Martin Buffery leading on interior. Talking to *Interior Motives* magazine in 2006, Buffery described the importance of the command driving position: "The mass of the interior is at hip level and it has a light, airy cabin feel. We're about the spirit of adventure and you can't do that if you can't see what you're having an adventure about."

The design team acknowledged that, while CB40 had succeeded in being regarded as a Land Rover, it lacked a premium feel. This was addressed with the new one-piece slush-moulded IP for a good fit and finish, as well as tactility. The new design built on themes from the L30 with its horizontal theme to the main IP and a strong vertical emphasis to the centre stack area with upright air vents.

In its second generation the Freelander matured into a more sophisticated product than the original. The spare tyre was relocated from the rear door to underneath the trunk floor, and the side-hinged door itself became a conventional tailgate, losing the electric drop glass function. Pushing the tracks out by around 100mm helped the stance and gave more interior space, too. There was an acceptance that the car needed to appeal to estate car and MPV owners, and the interior should feel familiar and user-friendly to these mainstream customers.

It also needed to appeal to another group of downsizers: "There are customers whose lifestyle no longer demands a vehicle as large as a Range Rover, but they don't want to give up the prestige of Land Rover ownership or a Range Rover's luxury. With the new Freelander we're offering that alternative," continued Buffery. Door ingress was improved over CB40 with less tumblehome and slightly larger door apertures, despite the stiffer structure to meet the latest NCAP side impact regulations. "The old car was quite pin-headed and claustrophobic, the new one has more upright side glass, better headroom and a more spacious cabin," he commented.

Cabin ergonomics followed the pattern set by the latest Discovery and Range Rover Sport, with the main lighting switch located on the end of the IP and window switches on the door belt line. The logical sequence in the centre stack of navigation screen, major switch pack, radio and HVAC controls was continued, while the control for the simplified Terrain Response system was located in front of the gearshifter. For the first time, serious consideration was paid to dog owners and how their pets might be accommodated, leading to the development of 'Roly' the digital Labrador to help with CAD packaging.

For the L359 project, the design process took a slightly different route from before. At an early stage in spring 2002 customer clinics were held around Europe, with the first proposals shown via animated CAD models developed using Alias software. The research focused on the various exterior theme directions, which ranged from very strong geometric product design-inspired proposals through to a mini Range Rover type of design direction.

Following this feedback, nine 40 per cent scale models were produced by June 2002, with three being directly milled out as 'speedfoam' models in polystyrene, and the other six generated traditionally in clay. Then three full-size clay models were developed, representing three directions:
- Premium Sports – by Earl Beckles
- Premium Utility – by James Watkins
- Sports Utility – by Sean Henstridge

In August Theme 1 was chosen, representing a good balance between premium and dynamic sports. The proportions were good, and it demonstrated a blend of the sheer surfaces of Discovery 3 with the expressive emotion of Range Rover Sport.

With the decision to base L359 on the EUCD platform at the end of that year, the approved model needed some adjustment, which took up the majority of 2003. In April 2003, a fully representative clinic model was developed for research in the US, a key market that the old model had never fully cracked. This model had an overtly Range Rover style of grille and front end that was later revised by introducing more vertical elements into the vehicle's nose.

By January 2004 the final design was frozen and detailed engineering work was now under way. "The exterior style was based on interlocking constructed components all coming together to form the signature lines of the vehicle," says Geoff Upex. "The form language is all about precision machined solid volumes of material." The idea was that the vehicle should express high quality engineering plus the enjoyment of a construction toy and inspiration was taken from chunky, solid kit such as bearing races, high end fishing reels and cycle gear clusters.

First prototypes were built in late 2003, just before the project was officially approved for production by Ford. These prototypes used CB40 bodies with new running gear so they could run relatively undisguised. Then 100 confirmation prototypes were built in early 2005 for main development testing using Ford and Volvo test facilities – one of the benefits of the PAG setup.

Under the L359 programme the three-door version was dropped. What had seemed to be a growing market had remained confined to the Japanese brands, and Land Rover planners decided to concentrate their efforts on the five-door market, in which sales volumes were more established and the Freelander could more easily maintain its market share. The launch for the Freelander 2 was at the London Motor show in July 2006, and the vehicle went on to be successful for over eight years, with production ending in December 2014.

Land Rover Design – 70 years of success

Cabin design followed the theme of the latest Discovery and Range Rover Sport, with a dominant centre stack containing the navigation screen, radio and HVAC controls. Launched as the Land Rover LR2 in the US market, L359 was a £600m investment by Ford, including upgrades to manufacturing facilities in Halewood.

The final Freelander 2 was a blend of the rational design of Discovery 3 with a hint of Range Rover Sport at the rear. The PSA-developed TD4 turbodiesel was the main engine, with 157bhp. A Volvo 3.2-litre transverse straight six was offered as a flagship engine, with 230bhp.

Land Rover Design – 70 years of success

Defender updates

As it passed into its sixth decade in production, the Defender continued to be manufactured at the rate of around 25,000 per year. Although it had never formed a large part of the design studio activities, the Defender was subject to regular engineering updates to keep it saleable, particularly with regard to engine emissions legislation, and limited design input was still required. This tended to focus on interior upgrades and regular freshening of paint colours and trim.

Defenders were still built using traditional methods on the line at Lode Lane.

In October 2001, the Defender was updated with zinc-plated steel side and rear doors for improved quality and finish. Whereas the tolerance on early Land Rover panels was +/- 0.25in (6.5mm), now it would be a more reasonable 0.2mm. This also permitted the fitting of electric windows and central locking for the first time on a Defender. To accommodate the electric window switches, the centre of the IP was revised and instrument illumination was improved.

At the same time, the 'XS' was introduced as a top specification level and the 'County' package could be applied to every model in the line-up. XS models came with many luxury features fitted to the Discovery, such as heated windscreen, air-conditioning, ABS and traction control. Heated seats and part-leather seat trim also became available.

The next major change was in late 2006 when the Puma 2.4-litre TDCi diesel engine from the Ford Transit was fitted, combined with the GFT MT 82 six-speed gearbox. This required a new radiator package, which meant the front grille would be pushed even further forward, protruding ahead of the headlamps. The new Ford engine led to further changes, recalls Peter Crowley: "The sump on the Puma engine couldn't be changed as it would no longer be a Ford-

Externally, the 2007 model year Defender could be distinguished by the pronounced hump in the bonnet, which was needed to clear the new Ford 2.4-litre Puma engine. A bold 'LAND ROVER' script replaced the small Defender badge used previously. To allow for the new radiator package, the front grille was pushed further forward for 2007, with a body-colour frame now surrounding the grille.

The Ford years

Land Rover Design – 70 years of success

The new IP introduced on the 2007 Defender – codenamed L316 – was a single-piece moulding incorporating round air vents in pods at the top that were sourced from the Ford Fiesta. "We saved £200 per car with this new IP," says Andy Wheel. This picture shows the 2015 Heritage Edition.

approved finished item so the engine was jacked up to give sufficient axle movement. That required a new bonnet. The higher gearbox position now required a new floor tunnel. That meant the current air con couldn't be fitted as the gearbox was too high. And so on …"

Thus, the opportunity was taken to fit a new ventilation system to improve demisting and heater performance, and this demanded a new full-width IP with face-level air vents. Here was an opportunity to finally ditch some ancient interior fittings that dated back to the 1970s, even if the steering column switchgear (sourced from the Austin Metro) and the ignition switch (from the Morris Marina) were still carried over.

The new IP design was based on a single, large moulding supported on a robust steel rail to help eliminate squeaks and rattles. Andy Wheel explains the story: "The Darwinian evolution of

The new IP and ventilation system necessitated the removal of the distinctive air vent flaps underneath the windscreen, which had been a feature of previous Land Rovers since the 1950s. While the flaps were deleted, the bulkhead pressing remained the same, so the outlines of where the flaps would be are still present. Note the bonnet bulge to clear the Puma engine. (Author's collection)

the Defender meant the IP had become a series of add-ons as the vehicle developed over the decades. We were able to consolidate it, make it more cost-effective, lighter, and simpler to assemble." The instruments came from the Discovery 3, and details such as all-LED illumination helped ensure high standards of reliability over rough roads. Versatile stowage shelves were conveniently located for both the driver and passenger and two console options were offered: a practical open-tray design that kept contents to hand or a large, lidded design that provided 14 litres of stowage space.

New EU4 standards demanded not only lower emissions, but reduced drive-by noise levels, too. According to Alan Mobberley, the sealing around the nose to reduce engine noise was so good the horn could no longer be heard – hence those three little vents between the side and turn signal lamps had to be added!

Land Rover Design – 70 years of success

Other interior changes were to the seating layout. EU legislation now outlawed the inward-facing seats used in the rear of previous Defender Station Wagons, and the 2007 model saw the four inward-facing seats replaced with two forward-facing seats. This made the Defender 90 a four-seater vehicle (reduced from six or seven), and the Defender 110 a seven-seater (reduced from nine).

A new bodystyle was introduced on the 110 chassis, too – the 'Utility.' This was a five-door body, but with the rearmost seats removed and the rear side panels left without windows, producing a five-seater vehicle with a secure, weatherproof load bed.

SVO and Special Edition Defenders

A total of 14 separate Defender body styles – from pick-ups and soft tops to crew cabs and station wagons – were produced on the mainstream production line at Solihull in 1999, and sold in over 140 countries. Beyond this, Land Rover's Special Vehicle Operations (SVO) team offered a range of products from standard drop-side or box-body conversions to bespoke design and build adaptations into ambulances, mobile hydraulic platforms or airport fire tenders – all of which were covered by Land Rover's warranty.

At the other extreme, standard models were available for commercial users, such as emergency services. The 130 was still available with the five-seater HCPU body as standard, too.

To maintain sales, Land Rover increasingly offered limited-edition variants of the Defender. In 1998, Land Rover offered the Defender 90 50th Anniversary model painted in Atlantis Blue micatallic. A total of 385 of these were built with a 4.0-litre petrol V8 and automatic gearbox. The specification included a half roll cage, stainless steel side runners and brush bar, spot lights, air-conditioning, and Raleigh cloth seating. Each had a numbered plaque on the right-hand rear panel.

Other Defender 90 special editions included the Tomb Raider in 2000, the X Tech in 2000-01 (based on the Hard Top) and the Hawaii, a Soft Top 90 model finished in Monte Carlo Blue and available only in France during 2001. The 150-strong Defender 90 Heritage Edition from August 1999 was reminiscent of an early Series 1. It had Atlantic Green paint, body colour alloy wheels, leather trim, special interior detailing and a wire mesh-type grille.

Defender 110 special editions included the Heritage Edition in 2000 (150 Station Wagons in Bronze Green with special grille, leather trim and other details) and the 110 Black (150 Double Cabs from June 2002, in gloss black with leather trim, checkerplate on the body panels, side runners and an integral roll cage). The Black was available as a 90, too, with 100 examples built for the UK from June 2002. Other special editions included the G4 Challenge in 2003, painted in orange.

A further extended Defender became available in autumn 2001. Land Rover South Africa needed a large capacity Station Wagon to meet the needs of safari tour companies, and in co-operation with SVO at Solihull developed a 37in chassis extension for the Defender 110. The Defender 147 was offered only as a six-door Station Wagon, and low-volume production was confined to South Africa. In Solihull, 16 147in chassis were built by SVO for use as plant tour buses.

Finally, a new SVX series was produced in 2008 as part of the 60th anniversary celebrations. This used the bug-eyed 'Mickey Mouse' look silver front grille and lamp surrounds that had originally been designed for the L316 updates, but rejected on cost grounds. Available in black as either a Soft Top or Station Wagon, 1800 examples were produced in total.

The magnificent Defender 147. The vehicle could be configured to seat 13 passengers, or with fewer seats and greater luxury.

The 2008 SVX was a limited edition of 300, in black and featuring Recaro seats in the cabin.

Chapter 7

Gerry McGovern takes charge

2007-2010

With the Discovery 3, Range Rover Sport and Freelander 2 all successfully launched, the designers in the Land Rover studio were looking forward to a period of consolidation and calm in the second half of the decade as part of the Ford empire. It was not to turn out that way.

When Alan Mulally became Ford's President and CEO in September 2006 he introduced a new 'One Ford' strategy that saw the company adopt a global product approach. As part of this implementation, it decided to divest itself of its luxury brands that made up Premier Automotive Group, starting with selling 85 per cent of Aston Martin in 2007 to a consortium of investors headed by Prodrive's David Richards. Aston Martin had recently established a new factory and headquarters at the north end of the Gaydon site and so now there would be upheaval and the facilities needed to be formally separated.

By this point Gerry McGovern had returned to London to run Ford's Ingeni 'urban think tank' studio. When this was axed, he had rejoined Land Rover as Director, Advanced Design in April 2004, but still retained his house in Chelsea, commuting out of the capital up the M40 to Gaydon. In August 2006 it was announced that he would replace Geoff Upex from the following January, and take charge of the whole design operation.

Upex explains his reasons to leave at that point. "I'd been in the role for eleven years, and had seen the complete regeneration of the Land Rover range. So, for me, it felt like now was the time to go." With the Freelander 2 just launched to great acclaim it would be going out on a high, but there were dark clouds on the horizon.

Ford sells to Tata

From mid-2007 Ford made it clear that JLR was up for sale, throwing the future of the whole company into doubt, not least for the designers and engineers based at Gaydon. The following months were a period of uncertainty and indecision, as programmes were put on hold, pending a decision on the company.

At the Geneva Motor Show in 2008, Ford announced that Tata Group of India was now the preferred bidder for the Jaguar and Land Rover brands. Tata Group was a fast-growing multinational conglomerate with deep pockets and a keen ambition to become a major player in the car industry. In 2007, Chairman Ratan Tata and Managing Director Ravi Kant had paid £6.7bn for Anglo-Dutch steel firm Corus, which was a major supplier to the industry, so it made sense for them to expand their portfolio in the UK, where they already owned over 18 businesses including Tetley Tea and the Courtyard 51 hotel in London. Progress was very swift, with the announcement on 27 March 2008 that the sale would be realised for £1.15bn, and the deal was completed by June.

Ratan Tata was reassuring and saw no problem when quizzed about Indian ownership of Jaguar and Land Rover: "We are very conscious of the fact that the brands belong to Britain and they will continue to be British. Who owns them is not as material as

Land Rover Design – 70 years of success

Gerry McGovern

Gerard 'Gerry' McGovern was born in Coventry in 1956, to Irish parents. Gerry was the youngest son in a family of boys, with brothers who were ten years older than him. As a teenager, Gerry was often taken to London by them, as they were already living the Swinging Sixties there, and this allowed Gerry to gain a love of that decade's culture that has remained ever since. He attended Binley Park secondary school, where an influential art teacher, Steve Chaplin, got him interested in car design as a potential career.

"Mr Chaplin had a relation that worked for Chrysler and when I was in sixth form he got me an interview there with Rex Fleming. I had all my art work. Fortuitously for me, Rex was ill and his boss Roy Axe saw me. I showed him my work and he suggested to go away and do more car sketches. So I came back, and he explained that he'd been struggling for years with training car designers. His proposal was he'd like to try to train a designer in-house with some training at college. Someone young with artistic talent who has cars in their blood. I was the guinea pig for that."

The meeting proved a turning point and Axe was to remain a major influence for McGovern. "Roy was an important figurehead in that he was one of the founding fathers of putting a professional design system in place, the types of skills needed, the disciplines and way you operate. He has written about how design didn't have any power, how design was treated as a service to the business. He went to America, brought those ideas back, tried to elevate the importance of design and I guess people like me have moved it on from there but, in fairness, he created a lot of that foundation."

McGovern started his studies at Lanchester Polytechnic – now Coventry University. He admits it was his training with the designers in the studio rather than his tutors that made him, and by age 19 he moved on to study at the RCA in London, sponsored by Chrysler. After graduating, he went over to Chrysler in Detroit in 1978 before returning to the UK in 1980.

When the Whitley studio closed in 1982, McGovern was soon snapped up by Axe to come and work up the road at Canley, where he led the design on the MG EX-E concept, and subsequently the MG F. That was followed up by CB40 Freelander project, following which McGovern was made Design Director for Land Rover in 1995.

Four years later Ford hired McGovern to head up and rejuvenate Lincoln-Mercury design. He set up a new design studio in Irvine, California and became only the second design professional to be appointed to the Lincoln-Mercury Board. Returning to the UK again in 2003 as Creative Director of Ingeni, Ford's design and creativity studio in London, McGovern rejoined Land Rover as Director, Advanced Design in April 2004. In January 2007 he replaced Geoff Upex as Design Director, becoming a member of the Land Rover board of management in 2008 and the Jaguar Land Rover executive committee the following year.

As the first designer to sit on Land Rover's board of directors, McGovern has been instrumental in promoting the importance of design to a brand that has traditionally been marketed on what it can do, rather than how it looks. "The Evoque has become a catalyst for change in our business," he said in 2011. "It represents the realisation that the design cannot be a consequence of the engineering or the manufacturing. Design has to be something that's at the very core of your brand and has to be given a very high priority. I think that design has to create an emotional connection."

McGovern received an honorary doctorate from Coventry University in April 2016.

Gerry McGovern.

the brands themselves, the enterprise and the people," he concluded at the press conference. The timing was fortuitous, despite the impending financial crisis later that year. The fact that several new models across the JLR empire were reaching the end of their development and were ready for launch – such as the Range Rover Evoque, Jaguar XF and XJ – would mean that Tata would subsequently see a swift return on its investment.

As Ford implemented its 'One Ford' plan, the selling off of assests continued. Following the sale of JLR, Volvo Cars was offered to Geely in 2010. The Mercury brand was also discontinued at that time, and Ford's 33 per cent stake in Mazda was sold off over a period from 2008-10 to stabilise its holdings. Over the past two decades Ford had spent $17 billion on building up PAG, but now, for the first time in many years, Ford management needed to focus entirely on its core products, utilising the Ford and Lincoln nameplates for all their markets around the world.

Richard Woolley had spent the period 2005-08 working on Ford programmes in the US, and arrived back at Gaydon in the middle of this turmoil. "Under Ford you were part of a big family, there was a lot of back-up there," he says. "Being sold to Tata there was a degree of trepidation. It was saying goodbye to yet another group of colleagues, just as we'd done before with BMW and Honda. Ford had to move very quickly to divest itself of those non-core businesses. It all happened very quickly. It's been a bit of a roller coaster ride!"

A new baby Land Rover – LRX

Thoughts about a more compact, lifestyle-orientated Land Rover had been on the drawing boards for some years, but were given new impetus with Gerry McGovern in charge. The design team had identified a 'white space' in the market. It was felt there was a market for a smaller, lighter coupé model with premium values, one that could give Land Rover a more upmarket feel. Audi had shown the Steppenwolf concept in 2001,

Gerry McGovern takes charge

Initial LRX sketches by Jez Waterman.

Jez Waterman.

but nobody had revisited this idea of a small two-door SUV coupé in the ensuing years.

First sketches for this new concept were done in spring 2006, and over the following 12 months the ideas were developed into a series of models. In an effort to get some cross-fertilisation of ideas across JLR, Julian Thomson was brought in from Jaguar design to lead on the project, supported by Jez Waterman for exterior design, Sandy Boyes and Mark Butler on interior, and Joanna Keatley for colour and trim.

Land Rover Design – 70 years of success

LRX was started in 2006 and was only Land Rover's second proper concept vehicle. Julian Thomson is seen here with the clay model.

Gerry McGovern takes charge

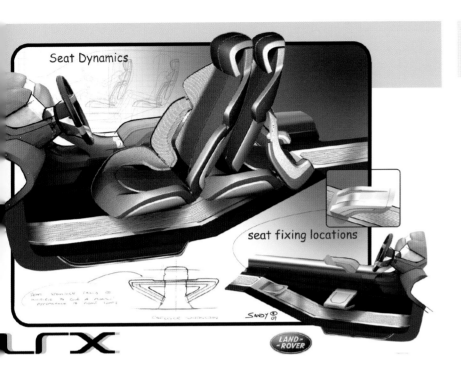
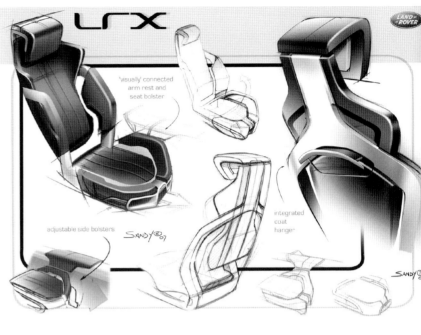
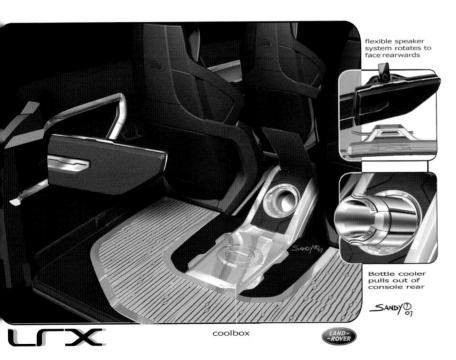

LRX interior sketches from 2007 by Sandy Boyes, showing new functionality. The aluminium seat frames could accommodate the front forks of two mountain bikes. The removed front wheels could then be stowed in dedicated slots in the trunk floor, which also had a coolbox embedded in it.

Land Rover Design – 70 years of success

16 models were initially made, and these were whittled down to a final three. "One concept had an incredibly sophisticated, premium look but looked a bit too 'today,'" recounted McGovern to *Interior Motives*. "Another design by Jez Waterman looked very youthful and dramatic, but didn't look Land Rover enough. And there was another that came from the school of functionality that – while I really liked it – didn't resonate with a lot of people because of its overt functionality."

Hence, the best features were combined into a second stage model that was refined into 2007. In summer that year the decision was taken to show the car – now called LRX – and work was started to build a running prototype, with the project done off-site at Concept Group International in Coventry.

The basic profile of the LRX was arresting. A steeply rising belt line was combined with a falling roofline and muscular wheelarches to give a powerful stance that was instantly appealing. The front mask introduced a new theme where the wide face was pulled around the corners of the car and right along the bodyside as far as the door, with the bold front wheelarches breaking through this graphic. Up close, the actual materials of this wide graphic cleverly shifted from being a grille texture, to a dark-framed headlamp, then an LED running lamp, next a fender vent and finally an aluminium insert in the door.

Interior design work started in 2007, and developed ideas shown on the Range Stormer. Mark Butler explains the concept: "You've got all this exposed structure across the facia. Every vehicle has a

Interior final rendering by Mark Butler. The aim was to show a lightweight, efficient interior.

Joanna Keatley developed a striking interior ambience based around soft tan and dark chocolate leathers offset against the pearl white exterior and polished aluminium details. Note the Lama carpet with embedded LEDs.

magnesium alloy cross-car beam that tends to get covered in layers of plastic, then we as designers have to fight like hell to get real metal back into the car. The thinking was to bring the components up to the surface and just apply local padding where you need it. We wanted to show how we could make a very lightweight, efficient interior."

The interior also showcased a new idea for an iPhone to dock with the centre console and act as the main control interface. "We don't have to carry around big in-car entertainment units, and it benefits the customer because the interface is known to them. When you dock the phone the start button appears, and when you press it all the phone's graphics appear on the screen behind the driver's wheel," says Butler.

Developing working screen graphics was an area that was still in its infancy in Gaydon studio and so London-based consultancy Imagination were brought on board to develop the ideas. Different layers could be switched off to personalise the display, while twisting the Terrain Response knob changed the ambient lighting to indicate the driving mode selected: green lighting for eco, blue for normal and red for sports – an idea later adopted by Mini!

The LRX also showed some neat ideas for flexible functionality. The console between the twin rear seats housed a bottle chiller that could be removed and slotted into the rear tailboard. Sandy Boyes designed the rear seats so they could slide forward to nest into the front seats, with the aluminium seat frames allowing the front forks of two mountain bikes to hook on. The removed front wheels could

Land Rover Design – 70 years of success

In a repeat of the success of the Range Stormer concept with an American audience, LRX was shown at Detroit Motor Show in January 2008. Lexan polycarbonate glazing was used for the side glass and the huge panorama roof. Note the keystone indexes in the wheelarch lips – a detail much copied since.

then be stowed in dedicated slots in the trunk floor. Boyes also designed a system whereby the Harmon Kardon rear speakers could be swung out and face rearwards onto the drop-down tailboard for impromptu parties, or removed altogether for parties further afield.

Joanna Keatley introduced a 'cell carpet' by Dutch design company Lama Concept, with embedded LEDs interspersed along the wool felt strips – something that had been shown for domestic interiors but had never been applied to a car interior.

By the time that LRX was showcased in January 2008, Ford was in the process of pitching the sale of Land Rover to potential bidders, and the show car took on a whole new significance as a demonstration of the design capability and potential for the brand.

Gerry McGovern takes charge

The LRX Design team at the Detroit Motor Show. (L-R) Sandy Boyes, Joanna Keatley, Gerry McGovern, Julian Thomson, Jez Waterman, Mark Butler. (Courtesy *Car Design News*)

In a bid to woo investors and indicate that Land Rover was taking sustainability seriously, the LRX was billed as a diesel hybrid 4x4 concept. A target emissions figure of 120g/km CO_2 was quoted to show that a vehicle like LRX could help Land Rover to avoid hefty CAFÉ penalties – something that investors would be keen to note.

This logo was designed as a shorthand symbol to show the arresting profile of LRX. The LRX was 150mm shorter and 205mm lower than a Freelander.

Land Rover Design – 70 years of success

The growth of digital design

Digital design methods have steadily replaced traditional paper drawings and physical models as the prime method to develop a car design. Oliver le Grice, James Watkins and Earl Beckles were part of the first group of designers in Rover to embrace digital design methods in the mid-1990s, initially using Adobe Photoshop, but later using 3D modelling software such as Autodesk Alias.

"At Gaydon I got totally immersed in the digital process on Alias, the speed of it, straight out of my head into 3-D," says le Grice. "It would take about two weeks for a model, then render it. I was one of four or five designers starting to get to grips with that. A great way of working, but a bit difficult, possibly opaque, for our managers, they never knew quite was going to come out."

The group pioneered several methods for milling from data, which have since become standard procedures. Milling of clay models was first done at Canley but they found it made a real mess, spitting clay everywhere. Milling polystyrene foam models in a dedicated workshop space was easier, sometimes using a thin PU skin to give a finished surface.

"After a couple of years at Gaydon we set up the big visualisation suite. The idea was to work on a PC as a designer looking directly at the full-size image on a huge screen. I guess I was at the forefront of development of those techniques. We developed 'Fast track to the Future' – a wholly digital workflow process. This was quite early days, way before this was seen as a viable method but we thought it would be. That was fascinating, not just sketching advanced design but developing advanced techniques for design."

Some years later, in July 2008, JLR invested £2m in a new 'Virtual Reality Centre' at GDEC. This state-of-the-art facility drew on advanced audio-visual technologies with the projection power of high resolution projectors to significantly speed up product development cycles.

The new Virtual Reality Centre allowed engineers and designers to see and interact with life-size, three-dimensional models of vehicles and components, reducing the need for physical prototypes. A total of eight Sony SRX-S105 ultra-high resolution projectors provided a visual quality that outstripped existing industry virtual reality Powerwalls and CAVES. The user wore 3-D glasses to experience 'ultimate realism' that simulated vehicle exteriors and interiors and was capable of making bodywork appear solid or transparent, all at a resolution that was near photo-realistic.

"That lead to animations. We started with two people, now there are teams of 15-20 pumping out film level quality material. I can't imagine doing without that now. Those have all become standard methods."

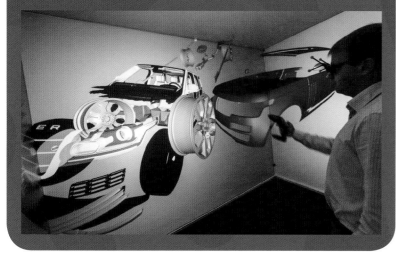

From LRX to Evoque

The LRX concept proved a big hit with the media, and there was an immediate interest from the public to produce such a car. As soon as Tata had signed the deal, Rattan Tata was keen to see the LRX in production, in both Coupé and five-door versions.

Andy Wheel worked with Dave Saddington as part of the team tasked with realising the concept into production and confirms that other voices within the company were less convinced. "There were a lot of naysayers," he says. "I remember one executive who said if this was a good idea, somebody else would have done it by now. The predicted sales volumes were pitiful. As we were developing that car the financial crisis hit. It was not a sure-fire money maker, there was always the sword of Damocles dangling over it. But there was so much belief in the car that we did find ways to make it add up from a financial point of view, trying to find ways to add more commonality, a real business-minded approach to protect the design."

One example was the roof panel, where the Coupé panel was rotated in space and commonised to give better head clearance for the five-door, which would be slightly taller. Even though that meant a marginally different side glass tumblehome to compensate, the cost savings in design time and engineering development were substantial.

McGovern was determined the production version should not be compromised. "I can remember one of the first discussions we had about this car. The engineers said, 'Right we'll need to put smaller wheels on it. We'll need to increase the gaps above the tyres to get a reasonable amount of wheel travel. It needs to sit higher. We're going to lift the bonnet by 90mm or more for crash testing.' And I said, 'Sorry but if we don't make the car like the concept we're not doing it.'"

In terms of the underlying architecture, Land Rover decided to continue to develop the LR-MS steel platform that underpinned the Freelander 2. The production LRX, codenamed L538, would share wheelbase and front bulkhead structure with that model, but 70 per cent of the underbody would be different. Tracks were pushed out around 20mm to approximate the stance of the concept, and the rear underbody was redesigned to achieve the low rear seating position to allow the dramatic falling roof profile. Just as important, there was a new front subframe that permitted the fitment of electric power steering, plus a lot of aluminium used in suspension components to take weight out of the vehicle, improve economy and reduce CO_2 emissions.

Diesel engines would be the same as used in Freelander but a new petrol engine was added, a 2.0-litre 237bhp (240ps) variant of Ford's turbocharged Ecoboost engine. Most versions would be 4WD but versions with just FWD would be available powered by the base TD4 diesel engine.

To ensure the production vehicle stayed faithful to LRX, McGovern insisted on the model being placed in the studio at all times, a point that Andy Wheel says made a huge difference. "That's the really

Evoque clay model in Gaydon studio. Considerable skill was required to translate the LRX design onto the D8 architecture.

important bit," says Wheel. "Most people who had seen LRX would have seen images in a magazine, a few might have seen it at a show. They would have had a perception of that vehicle based on fleeting imagery. Whereas in the studio we had LRX there next to the clay models. We were our own harshest critics, we had a perspective that no one else had. But that allowed us to go that extra mile."

"There were a whole load of attributes we had to deliver," he continues. "You don't design a show car to meet pedestrian safety standards, you don't have the timescale for that. The critical areas were the relationship of roofline to DLO, to shoulder to midriff features. Getting all that proportionally correct meant changing everything – the ripple effect. The trouble was it was reducing in scale width-wise yet the bonnet height needed to go up to meet pedestrian impact requirements for NCAP safety points. It was all going the wrong way – narrower and taller. So we had to critique everything to make sure we were getting the best balance of proportions so the overall impression out on road was 'Wow, that's the show car! And I can buy it.'"

As a proof of the result Wheel offers this anecdote: "I remember I was asked to show some VIPs around the Evoque. We had the master model of the car under a sheet. I whipped off the sheet and this guy said 'No Andy. I asked you to show us the production version please.' I said this is the production version. 'No this is the show car,' he said. 'No look, it's got door handles,' I said. That was the tick in the box for me."

In July 2010 Land Rover revealed the production car – named

Land Rover Design – 70 years of success

LRX (above) compared to the Evoque (left). The Evoque used a heavily-modified development of the EUCD platform, known as LR-MS/JLR D8. The faithful translation from show concept played a big part of the success of the Evoque.

The Evoque interior (left) compared to the LRX's. Andy Wheel: "Evoque was absolutely transformational. Being involved in that car was wonderful."

Evoque – at an exclusive celebrity-studded event at Kensington Palace in London, ahead of the car's official unveiling at the Paris Motor Show. Billed as a Range Rover (and not a Land Rover), the launch coincided with the 40th anniversary of the original Range Rover.

Talking to *Interior Motives*, Gerry McGovern said "It appeals to customers who haven't considered us before. It will appeal more to female buyers due to its compact, more comfortable package, and it's something they might feel more socially responsible about." As part of that slant towards female buyers, a celebrity tie-in was arranged to coincide with the Kensington Palace event. Victoria Beckham, former Spice Girl and fashion designer, was brought in to endorse the vehicle and give the impression that she had been involved in its design, with the title of Creative Design Executive.

This was far from the truth, of course, and was derided in the media. "No one believes Beckham will be wielding a Copic in anger anytime soon, yet she will apparently be spending 20 days at the design studio each year, helping the design team understand what a new type of Range Rover customer wants and potentially working on a series of special interior-focused projects," reported *Car Design News*. Nevertheless, in 2012 a 200-off 'Victoria Beckham Edition' of the Evoque was unveiled in Beijing, with satin matte paint finish, genuine rose gold details and mohair carpets. It retailed for £80,000.

At the 2010 Evoque launch, Land Rover's Managing Director Phil Popham said "Land Rover has a record of creating new market segments, or at least guiding them in new directions. Think back to the original Range Rover, the Freelander or Range Rover Sport. We're optimistic that Evoque can achieve the same." Indeed it did. Evoque went on to sell at a steady 120,000 per year – twice the originally planned volume.

Evoque also proved to be a game-changer for the design studio. "Once it came out, it was a huge success," says Wheel. "It swapped around the perception of design, it gave respect to design within Land Rover. I saw not only the transformation in the business, but the passion across the company to deliver that car. People said I don't need a car like this but crikey do I want one!"

Sorting out the brands

Following his appointment, Gerry McGovern was keen to bring new talent to his senior team. One designer who got the call was Phil Simmons, a McGovern protégé from the Lincoln days, who rejoined the studio from Ford in August 2007 as Chief Designer for Range Rover.

"When I rejoined there was some debate within the company as to whether Range Rover could become a separate brand entirely from Land Rover, because Discovery and Freelander said that on the bonnet. Yet there was always a Land Rover roundel on the grille of a Range Rover. At the time the whole design department was split with different teams doing Land Rover and Range Rover projects. I guess being parochial and

Developed from the 2012 concept, the Evoque convertible went on sale in spring 2016.

Land Rover Design – 70 years of success

responsible for Range Rover design, it made sense to me to allow Range Rover to be separate and move away from Land Rover, but there were always many points of view within the wider business."

Like Simmons, some felt that Land Rover should be allowed to go down a road of rugged practicality and rational design, following the direction taken with L319 Discovery and Freelander 2, allowing Range Rover to pursue a more upmarket design language. Others, including McGovern, felt that the functional design ethos was too limiting, and there was increasing evidence that it alienated some customers and was holding back sales. In his opinion both brands needed to become more premium, with Evoque being a template for a more sophisticated design approach.

Despite the decision to put LRX into production at Halewood, things were not looking good for Lode Lane. At the height of the recession in September 2009, JLR announced that it would close either Lode Lane or Castle Bromwich, as both plants were operating at just half capacity. The Solihull factory had long been seen as the weakest link at JLR, the 308-acre site being hemmed in on all sides by housing, with little room for expansion.

McGovern made a strong case to push for rapid face-lifting of the range to improve sales, to add more glitz to the front of the cars and to improve the perceived quality of interiors, which were considered to be dropping behind the competition in terms of material quality and showroom appeal. Hence, the revision for the Discovery, when it became known as Discovery 4, plus the Range Rover Sport face-lift, known as L420, both done for 2010 model year.

If Spen King is regarded as the father of the Range Rover, then Phil Simmons is certainly today's 'Mr Range Rover.' Simmons has been involved in every model since P38A.

The Discovery 4 was an example of the push to add more glitz to the range and improve sales, a strategy that worked. For the 2014 face-lift, the name 'Discovery' and not Land Rover was prominently displayed on the bonnet.

Land Rover Design – 70 years of success

Dr Ralf Speth was appointed CEO of JLR in 2010 to stabilise the future of the company and combat the weak sales situation.

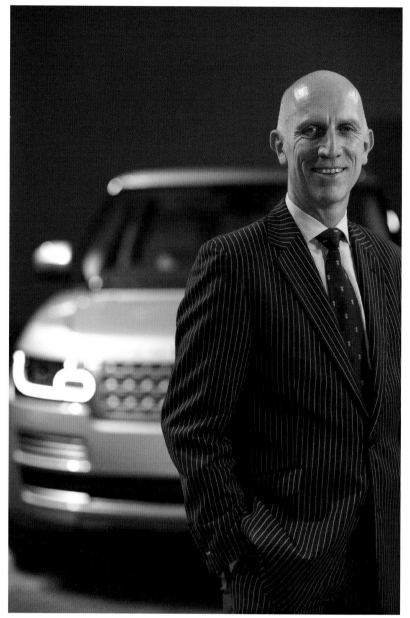

John Edwards, Global Brand Director for Land Rover 2010-14. Formerly MD of Land Rover UK, he brought a clarity to the debate on Land Rover brand.

"This is just part of the cultural change we're going through," said McGovern to *Autocar* in 2010 during the Evoque launch. "We need to design and build cars that people want, as opposed to them buying vehicles that we're capable of producing."

With the face-lifted models for 2010, sales picked up strongly and the factory closure threat receded. McGovern was aided by his new boss. Dr Ralf Speth was brought in as CEO of JLR in February 2010. A close associate of Dr Wolfgang Reitzle, Speth had been involved before with Land Rover, initially under BMW and later with PAG before the division was dissolved under Ford. Having worked at Linde for a few years, Speth was a natural choice for this new role, with a deep knowledge of Land Rover going back over 15 years, and was a strong ally for the Tata management, able to ensure that the falling sales of JLR could be turned around. Another ally was John Edwards, who was promoted to become Global Brand Director for Land Rover in December that year.

Simmons credits Edwards with bringing clearness to the brand debate, too. "One of the first things John did was to stop any further

LRX and 'Huey,' the the first pilot build Land Rover, at Lode Lane. Ironically, when this photo was taken in 2009 the plant was under threat of closure.

conversation on that side of things. He said Land Rover brand is the master brand, and includes Range Rover and Land Rover attributes. Thus the current situation has the sub brand clearly stated on the bonnet: Range Rover, Discovery, or Defender, whereas the master brand is always represented with the Land Rover oval. Under each sub-brand there are other model names eg for Range Rover there is Evoque, Sport or Velar. It finally makes sense when you see the three product families now. The name on the bonnet, and Land Rover master brand giving the promise of capability and off-road performance. John really brought clarity to that whole thinking."

"The conversation has moved on now, we have much clearer understanding of how sub-brands relate to master brand, which acts as a halo," he continues. "Rolled up in Land Rover are all those fantastic associations about our heritage right back to Huey and 1948. It's an absolutely crystal clear focus on form follows function, with fantastic off-road capability. The way it looks is a natural consequence of the way it is as a vehicle. That's a priceless heritage we're very fortunate to have. It applies to Range Rover as much as Land Rover. I'm very comfortable with the final decision we've made on that subject and it's no longer debated."

The turnaround in fortunes was helped by the sales success with the Evoque Coupé and five-door. This was boosted by the arrival of the convertible version, first shown as a concept in 2012. Simmons took charge of the production design. "The Evoque Convertible offers the best of both worlds: the crisp, precisely defined contours of the Coupé's striking silhouette with the roof closed and the excitement of top-down motoring with it open," he commented at the launch.

2010-2016

Chapter 8

Rapid expansion of design activities

Once sales began to pick up in 2010, JLR pushed ahead with plans to expand the portfolio of products it would offer. Ratan Tata and Ravi Kant remained true to their word, and continued to plough investment into the company, despite the global slowdown of 2009 that saw a downturn in sales and brought other carmakers close to collapse. Land Rover in particular was seen as a brand that should capitalise on its image, with sales of luxury SUVs growing strongly worldwide. Range Rover Sport had proved a huge success since its launch in 2005, and it was important to maintain the sales momentum it had created with timely updates and an eventual successor within its planned eight-year lifespan.

With Dr Ralf Speth heading up JLR and John Edwards in charge of Land Rover, a big push for rapid product expansion took place. On the design side, the number of designers involved in each project and their roles started to change. Traditionally, younger designers were tasked with generating more advanced concepts and initial ideas requiring a freshness of outlook that sometimes eluded more experienced designers, who were more concerned with developing resolved solutions that worked.

The hierarchy of design also meant that as staff were promoted, the role shifted to becoming a manager or a chief designer, with less time spent actually sketching or designing and more time spent managing teams and projects. Now, experienced designers wanting to concentrate on their artistic talent were termed 'creative design specialists' and able to continue up the career ladder using their core skills, just as happens for a physician or dentist.

Meanwhile, designers whose skills lay in refining and honing of ideas were encouraged to become 'realisation designers,' tasked with bringing a concept through the production development process, with its myriad of regulations and manufacturing constraints. Finally, as vehicles and programmes became more complex, a new generation of highly design-literate managers were needed to ensure the design intent was not lost as the vehicle was taken into production. Exterior and interior design teams were thus organised into 'Creative Design' and 'Realisation Design,' with separate directors for each team.

Two new Range Rovers in six months
The L322 Range Rover had proved a turning point in Land Rover's fortunes. Despite the recession, sales were running at around 35,000 per year, and it had regained Range Rover's image as the premium luxury 4x4 at the top of the market. Furthermore, it had given the design team a new confidence about the identity of Range Rover, establishing a design language that was uniquely British, something unmistakable yet understated that set it apart from its competitors. Planning its successor, codenamed L405, would be vital to get right – could it be improved upon?

Indeed, the initial design proposals for its replacement ran into problems. Richard Woolley was promoted to Advanced Design Studio Director to lead on the programme in 2008. "On my return from Ford

Rapid expansion of design activities

in Dearborn Gerry asked me to look after the new L405 from inception to production," he explains. "Work on L405 started in 2009. It had started the year before but there was some unease about the direction it was taking, so Gerry asked me to take a fresh look at it and produce an alternative design, which became the production car."

There was consternation within Engineering about the timing for the project with this reappraisal of the design direction for the vehicle. Up to that point a number of scale models and several full-size clay models had been done, with one selected to go ahead. With this shelved, Woolley's team needed to proceed quickly, and moved ahead with one new clay – termed a 'vision model.' This was approved and they then proceeded straight onto the production full-size clay model.

The Range Rover Sport project (L494) was considered at the same time, but needed to run slightly behind the main project to allow it to progress unhindered. This time around, the plan was for the Sport to use the same platform, termed D7u, rather than being based on the Discovery as before. L494 would now share the same 2922mm (115in) wheelbase as the latest Range Rover, providing more interior space with 118mm more rear legroom than the old Sport model.

Whereas L322 had used a mainly steel bodyshell with some aluminium skin panels, D7u was an all-new aluminium architecture, with the body a unibody as before. The main aim was to drastically reduce the weight of the vehicle so that less thirsty engines could be used without diminishing the formidable performance of the Range Rover. Car manufacturers were now faced with CO_2 targets for average emissions, and JLR was in a vulnerable position unless it moved swiftly to more fuel-efficient products.

Jaguar had pioneered the use of riveted and bonded aluminium bodyshells with the X350 XJ in 2001, so there was considerable expertise within JLR that could be brought to bear with this new project. Of course, when the previous L322 was being planned under BMW, Jaguar was still a separate

Four generations of Range Rover. At front is an Autobiography long-wheelbase L405 with Narvik Black contrast roof, panorama glass sunroof, and 21in 'Style 7006' alloy wheels. Noticeable here is the more rounded and raked front end compare to the previous L322 behind.

Richard Woolley was in charge of the fourth-generation L405 Range Rover, launched at the Paris Motor Show in September 2012.

Land Rover Design – 70 years of success

L405 Range Rover had enhanced capability over outgoing L322. Note the high-mounted engine air intake in the fender and the massive B-pillar of the lightweight bodyshell.

Ford-owned company, but now the two were united and the obvious weight-saving gains could be utilised for this programme.

The all-aluminium construction gave a 39 per cent lighter bodyshell, enabling total vehicle weight savings of up to 420kg. A new alloy, AC600, was developed for the exterior skin panels, allowing greater formability and tightness of feature lines and curve radii than was possible before. A second 6000-series alloy, AC300, was used for much of the body structure, while the one-piece bodyside was the largest single pressing in any vehicle, yet weighed just 7kg. From being under threat of closure, now £370m was invested in the Solihull plant to create the world's largest aluminium body shop.

The body was produced using no welds, but aircraft grade adhesives and 3400 Henrob self-piercing rivets, as on the Jaguar. In another bid for lightness, magnesium was employed for the front crush structure and SMC plastic used for the rear tailgate. Despite the weight reduction aims, a large panorama sliding glass roof option was developed, providing a light and airy feel to the cabin.

At just under five metres long, the new model occupied a similar footprint to the old L322, but with a smoother and more aerodynamic profile. New interpretations of the classic design cues were sought, including the clamshell bonnet and floating roof. The graphic of the side fender vent was now executed on the surface of the front doors. While some bemoaned the fact that the vent was no longer functional – simply a decorative motif – others admired the

Rapid expansion of design activities

The door side vent moulding could be ordered with a silver finish to match the lower bodyside moulding, giving a distinctive signature to the Range Rover.

With its more cab-rearwards stance, large side vents and pronounced taper towards the rear, the L405 design harked back to the early Riva theme pioneered by Phil Simmons for the previous L322.

Land Rover Design – 70 years of success

The interior was a cleaned-up version of the previous theme, with a broad beam that ran across the cabin, overlapping the centre console and distinctive wood cappings at either end containing the vents. The interior design was signed off in late 2009.

The Executive Class seating package included a deep console between the rear pair of seats. Three levels of Meridian sound systems could be specified, together with displays in screens in the front seat headrests.

way it graphically linked to a bright lower bodyside moulding that ran through and linked to the rear tail lamps. Similarly, while the belt line was still mercifully flat, the slight rise in its position shifted the command driving position closer to that of the Range Rover Sport.

"The exterior design of L405 deliberately avoided the over-complex surfacing that was becoming the norm in luxury SUVs, particularly in the bodyside section, bonnet and front bumper," explains Woolley. The front face gained a new benign expression that was imperious without appearing overly aggressive, with headlamp 'ears' that echoed the theme of the Evoque, and wide lower bumper graphics that enhanced the calm, assured look of the vehicle.

Rapid expansion of design activities

The visual centre of gravity was shifted on the new vehicle, too. Whereas the old Range Rover greenhouse sat squarely across its wheelbase, the new one was more cab-rearwards. This visual trick was a combination of slightly more body behind the rear wheel and 'faster,' more sloping A- and D-pillars. But the critical visual key was the much faster C-pillar, which now shot behind the rear wheel rather than visually indexing with its centre, as on L322.

A team of four designers under interior design manager Nic Finney developed the interior. Mark Butler was the lead creative specialist: "The Range Rover interior is very much about letting the horizontal features run full width, but encapsulating them with the end caps to frame the instrument panel. In the last-generation, model, veneer became a very structural component. We wanted to develop that," he explained to *Car Design News*.

By June 2009 the main theme for the IP was established, with a broad leather-trimmed beam running across the vehicle and over the centre stack. The veneered section of the centre console was deliberately made to run under the centre stack to enhance the impression of this vertical area being supported by the side stanchions, trimmed in aluminium. The switchgear count was reduced by over 50 per cent, with many functions consigned to the 8in infotainment screen. "When you first get into the car, the icons and everything else is hidden," says Butler. "They are all touch-sensitive switches … so the interface is clean and elegant. Then, when you power it up, it all comes to life."

Joanne Slater was the Chief Designer for colour and materials on the project. For the first time, material quality in the rear of the cabin was comparable to the front, and a broad centre console was fitted in the top-level Executive Class seating package, making the vehicle a strict four-seater. "You can see the two consoles, front and rear together, which really emphasises how we wanted the material quality to flow through the cabin." Rear controls for climate control

The sumptuous rear cabin of the 2014 LWB Autobiography Range Rover, with the longer console and extended door linings. It offers the experience of a private jet, including powered reclining seats and ottoman footrest.

Standard wheelbase L405 could be ordered with a body colour roof, or contrast black or silver roof. The dark example has a silver side vent and 22in 'Style 5004' alloy wheels.

Land Rover Design – 70 years of success

Side view comparison of the L405 Range Rover and L494 Range Rover Sport.

Rapid expansion of design activities

The L405 and L494 were designed in parallel, with the first sketches commencing in 2009.

and seat heating were built into the rear console, together with a lined storage box. A choice of three sound systems by Meridian was offered, including the Signature reference system – the world's first 3D in-car surround sound setup. A wider range of interior and exterior colour options was developed, including the option of a contrast black gloss roof colour.

A choice of two petrol 5.0-litre V8 engines were offered, with 375ps or a full-house 510ps supercharged version, with a top speed of 140mph. In addition, a 4.4-litre 339ps SDV8 diesel and a new 3.0-litre 258ps TDV6 diesel were available, the latter with CO_2 emissions of under 200g/km. To give the new model new levels of all-terrain capability, a second generation Terrain Response was developed that used intelligent systems to automatically detect the driving conditions and select the appropriate terrain programme. As before, air suspension offering variable ride height was employed, with axles cross-linked for maximum axle articulation. Revised routing for the engine air intake to the top of the fenders now meant the wading depth was 900mm, more than 200mm more than before.

New driver aids included Adaptive Cruise Control with Queue Assist that could bring the vehicle to a halt from motorway speeds, and Blind Spot Monitoring. This detects vehicles within a five metre zone to the side and 73 metres to the rear, and alerts the driver via an amber flashing lamp in the appropriate side mirror.

Woolley was also involved in the sister Range Rover Sport

Land Rover Design – 70 years of success

The Sport has more belt line wedge and roof taper than the Range Rover. The main recognition point from the rear is the Evoque-style horizontal rear lamps, rather than a vertical stack as on the Range Rover. The sketch is by Matt Dillon.

development, with the production launch following just six months later at the New York Motor Show in March 2013. Despite the reservations at launch in 2005, the first Sport had successfully added around 50,000 vehicles per year to Land Rover's production volumes and had greatly helped JLR's profitability. Having established itself as a key model in the line-up, the new L494 Sport was deliberately positioned at the centre of the three-model range, between Evoque and the full-fat Range Rover.

With its increased wheelbase, the packaging of L494 was tweaked to give a new twist to the formula and to further add to its appeal – the provision of a pair of occasional seats in a third row. These were deliberately not as generous as those in the Discovery, but were targeted towards children up to age of 12, where the restricted legroom was not so critical and the sleek roof profile of the Sport could be preserved.

Visually it also sat between its two siblings – less cartoonishly rakish in profile than the Evoque, yet with more belt line wedge and roof taper than the Range Rover. The graphics followed the themes

Rapid expansion of design activities

introduced on the other Range Rovers, with tail- and headlamps that bled into the fenders, and a new take on the fender air outlet. Here it was accentuated by a feature line that flicked back into the door surface. The deep bodyside was partially alleviated by a light catcher half way down the doors, although models featuring painted lower body mouldings can appear quite heavy-looking, something that also afflicted the Evoque.

The interior continued the Sports Command driving position of the L320 Sport, with a more reclined driving position than L405. The IP shared most of the L405 architecture, but added a one-piece pad across the cabin and a unique centre console with a stick shifter for the automatic gearbox rather than a rotary controller.

The engine line-up for the Sport was similar to that of the Range Rover, except a new Si6 3.0-litre supercharged V6 petrol engine with 340ps replaced the 375ps V8, giving much improved fuel economy and lower CO_2 emissions. In addition, a more powerful 3.0-litre SDV6 turbodiesel was added with 292ps and a 0-62mph time of 7.2 seconds – the same as the new petrol V6. The big news was the latest diesel hybrid powertrain, with low CO_2 emissions of 169g/km. This used a 3.0-litre V6 292ps diesel plus 35kW motor to give a total 340ps output. Both this and the new Si6 petrol engine were offered in the Range Rover from 2014.

Further derivatives soon followed as the pace of design and development at Land Rover was cranked up. A LWB Range Rover was introduced in 2013, with a 200mm increase in wheelbase to 3122mm, taking Range Rover into higher territory as a truly luxurious limousine to compete with the latest full-size SUVs from Porsche and Bentley.

From August 2014, UK customers could order the LWB Range Rover in Autobiography Black specification. This offered higher levels of luxury with an exclusive new Lunar/Cirrus interior colour combination, semi-aniline leather seats with precision detailing, and new features such as integrated USB

The new L494 has proved an even bigger success than the original Sport, currently selling over 85,000 units per year. A contrast roof colour was optional on the new Sport.

This final rendering of the L494 Sport IP shows how it shared much of the L405 interior architecture, but added a one-piece pad across the IP and a more raked centre console with a stick shifter for the automatic gearbox, rather than a rotary controller. The interior design was signed off in summer 2010.

Land Rover Design – 70 years of success

Occasional third row seats were a new feature for the L494 Sport, offering a more premium seven-seat vehicle than the Discovery.

charging sockets, enhanced stowage and bespoke lighting. The Executive Class seating package was enhanced, offering two individual fully adjustable seats in an extended centre console with electrically deployable leather tables.

In the same month the Range Rover Sport SVR was announced by Land Rover. The biggest change was the 5.0-litre supercharged V8 from the Jaguar F-type with a power output of 550ps, stiffened suspension and an active exhaust system with electronically-controlled valves. The improved performance made the SVR capable of 0-60 mph in 4.5 seconds, and gave a top speed of 162mph.

SVR design changes comprised 21in alloy rims, a new front bumper with larger side ducts, and the main grille finished in black. The rear bumper was also completely redesigned with a more pronounced diffuser and twin-circle exhaust finishers.

Discovery Sport

How to replace the Freelander was a major preoccupation for JLR after the split from Ford in 2008. Developing an all-new compact platform might be beyond the resources of the company that was now investing heavily in the larger D7u architecture for the Range Rover and Discovery. Therefore, it continued to develop the EUCD platform with its transverse engine layout, deriving the heavily

Rapid expansion of design activities

modified LR-MS architecture from it, also known as the JLR D8 platform. Variations of the D8 would be used for the 2011 Range Rover Evoque, and, later, the 2017 Jaguar E-Pace.

The Freelander 2 had performed well since its launch in 2006, but it suffered from a somewhat-diluted image identity. It was marketed as the LR2 model in the US market, and, following the debate about branding, there was agreement to move to a more unified global identity based around the core brand names of Range Rover, Defender and Discovery. Just as the Range Rover Sport was a clear derivative of its larger brother, could the new Freelander not be repositioned as more compact version of the well-regarded Discovery?

Furthermore, the market was changing. With so many compact SUVs flooding the market, the Freelander was in danger of getting lost amongst a sea of new models, particularly from Asia. Although the CB40 had been one of the pioneers of this segment in Europe, many others had entered the scene now, with far larger dealer networks and advertising budgets to boot. Maybe it was time for a rethink.

One of the core attributes of the L319 Discovery was its seven-seat layout that offered a wonderfully versatile and spacious cabin for wealthy middle-class families. Lower down the scale, families requiring more than three seats for children had little choice but to choose a monobox MPV, such as the GM Zafira, VW Touran or Ford S-Max. Gaydon's planners decided that a more premium brand such as Land Rover could offer a highly attractive SUV alternative that would be more unique in the market and provide a strengthening of the Discovery nameplate to boot. Thus, the L550 Discovery Sport project began to take shape.

The plan was for it to share most of the front of its D8 platform with the Evoque, but the rear end was completely re-engineered once again to make room for the third row of seats that folded into the floor. The wheelbase was extended by 80mm to 2741mm (108in) and a new multi-link independent rear suspension was part of the package to allow a higher payload that seven passengers and their luggage would demand. The occasional rear seat design – perfect for pre-teenage children – was modelled on that of the latest Range Rover Sport, with some components being shared. Second row seats could slide fore and aft to provide a variable package, with maximum legroom for lanky teenagers or reduced legroom to allow space for third row siblings.

In some respects it was a return to the original Pathfinder/Oden idea of a versatile compact people carrier, but updated to fit within an SUV format that would allow it to compete with the Volvo XC60 and BMW X3. The engine line-up followed that of the Evoque: the 2.0-litre petrol engine with 240ps or the 190ps SD4 diesel, shared with PSA and Ford. A 2.2-litre Td4 diesel with 150ps was later made available, too.

Ten exterior model proposals were developed, and by summer 2011 one exterior design theme had been chosen. However, over

---

Vision Discovery concept

The Vision Discovery concept was revealed alongside the Virgin Galactic spacecraft in April 2014 at a glitzy event onboard the US aircraft carrier Intrepid, in New York. Land Rover had just announced a sponsorship tie-up with the commercial space venture to ferry astronauts and customers to and from its desert launch pad, and this concept was part of the New York Motor Show activities.

Designed in conjunction with the production new Discovery, the project was done under Richard Woolley's direction, with Massimo Frascella in charge of exterior and Dave Saddington for interior. The concept showcased the essence of the forthcoming Discovery design language, with its softer and more sophisticated surfacing. The big change was the rising belt line and very fast C-pillar that pushed the visual centre of gravity quite high. The stepped roof was reduced to a softened bump in the roof and the Alpine windows were deleted. The rear was the most controversial area, with a pronounced chamfer to the rear corner and a very deep rear fender that made the rear look tall and narrow.

While the exterior was a faithful preview of the Discovery, the interior was deliberately more conceptual, although it still previewed the versatile seven-seat interior layout that would become a brand feature for the Discovery family of vehicles. Entry was via pair of barn doors with no central pillar. White leather as used on yachts and supplied by Italian supplier Foglizzo was used for the front two rows of seats, while dark blue nubuck leather graced the third row pair. The nubuck leather was treated to repel and protect against oil and water spillages, thereby making the vehicle "premium but not too precious," says Chief Designer for colour and materials Amy Frascella.

The Vision Discovery concept was first shown at New York Motor Show in 2014. For the Beijing Auto Show a couple of months later the model was resprayed Taklamakam Orange.

Land Rover Design – 70 years of success

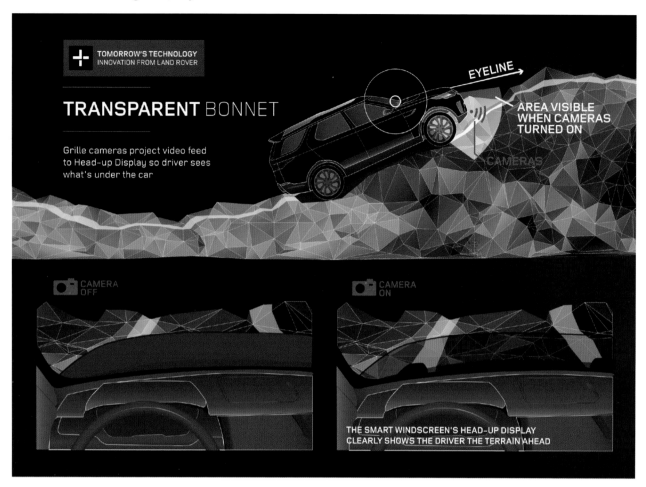

The Vision Discovery also showcased under-car camera technology that relayed images to driver screens, making the hood 'virtual see-through' while off-roading.

L550 Discovery Sport clay model in Gaydon studio. Phil Brown works on the rear.

The interior of the Discovery Sport was a scaled-down preview of the forthcoming new Discovery, with a return to the strong vertical stanchions either side of the centre stack.

Rapid expansion of design activities

Discovery Sport HSE 'e-Capability' in Indus Silver with 18in Aero alloy wheels. (Author's collection)

Wheelbase was extended by 80mm over the Evoque. The rear tail lamps with the bisected circle motif are a recognition point.

the coming months it was heavily revised as the Vision Discovery concept and New Discovery were developed, employing the new surface language that would be used across the whole Discovery family.

The strategy proved highly successful. By 2017 the Discovery Sport had become Land Rover's best-seller, making up 126,078, or 28 per cent, of all sales, while the Range Rover Evoque and Range Rover Sport amounted to 190,000 sales – a remarkable achievement for two models that suffered widespread scepticism within the company before being launched. Production at Lode Lane was at record levels and, far from being under threat of closure, the plant was now operating at maximum capacity.

New Discovery L462

"Don't get me wrong, I really like the last Discovery but [that kind of design] isn't appropriate anymore. It would have held us back," commented McGovern to CAR magazine at the launch of the new Discovery at the Paris Motor Show in 2016. "Some people – myself and other product designers

Dave Saddington grew up in Walsall and joined Austin-Rover in September 1983. He was Studio Director for the interior of 2014 Vision Discovery and Discovery Sport.

Land Rover Design – 70 years of success

(Left to right) Massimo Frascella, Gerry McGovern and Phil Simmons.

Five generations of Discovery. New Discovery was done under Richard Woolley's direction, with Massimo Frascella in charge of exterior and Mark Butler for interior design.

included – liked the design but it is alienating. The wheelarches for example, that constructivist design: they aren't modern anymore, they were of their era."

Replacing the Discovery was a vital project for Land Rover to get correct. Not only the design but the heavy weight and uncompetitive aerodynamics were major targets to be improved if the vehicle was going to continue its sales success into the 2020s. Use of the D7u aluminium architecture would drop the weight by up to 480kg, while the 0.4Cd drag coefficient of the Discovery 4 was given a target of around 0.33Cd for the new project, codenamed L462. The generous dimensions of the D7u architecture meant the wheelbase could now be common with the large Range Rover and Range Rover Sport at 2922mm (115in), a 38mm increase over the outgoing model, and sufficient to ensure really good packaging for seven adults. Overall length was up by 141mm (5.5in), but width and height were both slightly decreased to optimise the frontal area and help with lowering fuel consumption and emissions.

Design work started in 2012 with discussions amongst the team and within product development about the direction the new Discovery should take. "It's absolutely a two-way process with the engineers contributing as many ideas as the design team," explains Mark Butler, now Interior Creative Director. "It's really important they are engaged early on or there's no depth to the ideas."

Rapid expansion of design activities

Land Rover Design – 70 years of success

The launch in 2016 stressed the advances in versatility, capability and connectivity over the outgoing Discovery 4.

Rapid expansion of design activities

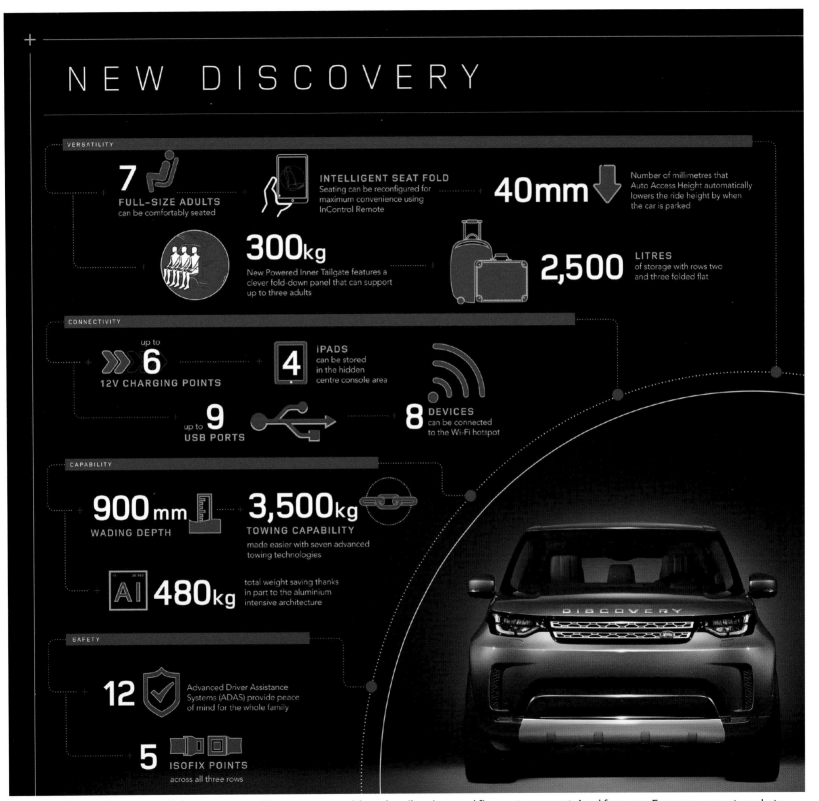

UK market vehicles were all air sprung and with seven seats, although coil springs and five seats were retained for some European export markets.

Land Rover Design – 70 years of success

Never had the studio seen so many projects under way at one time. L405 and L494 Range Rovers were deep in their final stages before pre-production, with many design refinements still ongoing. L550 Discovery Sport had just been started and needed a large team for exterior and interior work. Evoque was only just launched and the convertible version was still being developed. Meanwhile, there was another face-lift programme for 2014 for Discovery 4 plus the DC100 show car to take care of ... With so many projects under way at this point, the design studio organisation needed to be revised with a new Exterior Chief Designer being brought in to oversee this latest project.

Massimo Frascella was another trusted colleague of Gerry McGovern's from his Lincoln days at Ford. He had originally worked for Bertone in Turin before joining McGovern in England at the first Lincoln studio, based at TWR. The Lincoln team briefly moved into a corner of Gaydon studio before moving to sunnier climes in California in 2000. With the disintegration of the PAG setup, Frascella briefly moved back to London with McGovern but was then offered a senior position with Hyundai-Kia back in California in 2004. In summer 2011 he got the call from McGovern to join him again and take a hands-on lead for exteriors, starting with Discovery Sport, but more importantly with L462 programme.

"A few scale models were done initially, then Gerry made a selection of the theme that I made," says Frascella. This was then taken to full-size clay model stage towards the end of 2012. A see-through 'Design Vision' model was then made in summer 2013 to get the rest of the company engaged with the design direction and ensure that the spirit of the theme was maintained, as it was passed from the creation team to the second stage of development under the design realisation team.

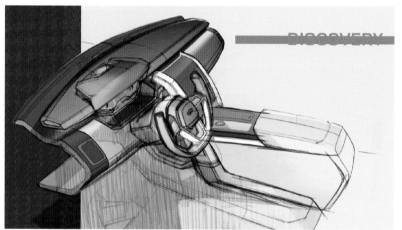 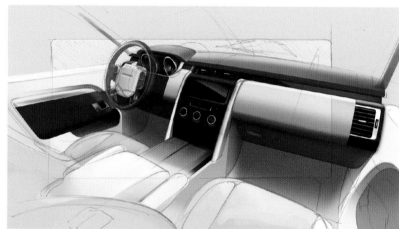

Early sketches for interior of L462 Discovery.

A secret storage pocket for phones and sunglasses was incorporated behind the drop-down HVAC control panel.

Rendering of L462 interior. Second row seats can slide 160mm fore and aft. The basic interior colour is Ebony but no less than four accent colourways were offered at launch: Acorn, Nimbus, Vintage Tan and Glacier.

Rapid expansion of design activities

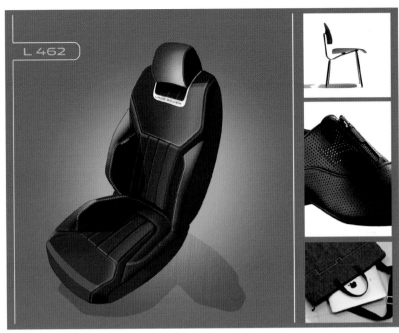

Seat sketch and mood board showing the influence of furniture and leather detailing from shoes.

Mark Butler was in charge of interior design, the second Discovery project he had worked on. "Discovery IP has a strong vertical overlapping the horizontal beam," he explained to *Interior Motives* magazine. "We put the switchgear into the primary zones, the rest of the cabin is about giving the customer the maximum versatility to use it as he/she sees fit."

One big change was the decision to ditch the asymmetric split tailgate of the outgoing model in favour of a huge one-piece sheet-moulded plastic tailgate. Discussions were full and frank, according to Butler. "There were advantages, but on a day-to-day basis there are issues. From an exterior point of view, it limits the amount of plan shape you can have at the rear; there are reach issues and there's not much protection from the weather. The focus for us, then, was how we could bring some of the benefits of the split tailgate back in." The answer was a new powered inner tailgate flap that gave the freedom for the exterior plan shape whilst providing some degree of load retention and reachability into the trunk. It also retained the perch seat function and provided a step for dogs to jump onto to aid getting in.

Three clay interior models were commenced in 2014. Two were closely related but the third was more different "To help us understand the boundaries," says Butler. One big step forward was to provide power folding for all the second and third row seats to ease the task of reconfiguring the seating for differing conditions. To reflect the increasing demands of children, the cabin could accommodate nine USB ports, six 12-volt chargers and a wi-fi hotspot for up to eight devices. New storage spaces were designed for drinks bottles, games and tablets and a neat sunglasses storage pocket was designed behind the drop-down HVAC control panel.

In autumn 2013 the decision was taken to present a concept version of the new Discovery family look with a show car. This was the Discovery Vision concept (see sidebar). By this stage the production Discovery exterior had been frozen and the show car was kept as close as possible in feeling in order to benefit from the same positive associations that made the Evoque such a success after showcasing the LRX concept.

Another new team member was Amy Frascella, wife of Massimo, who joined as Chief Designer for colour and materials. The importance of more colour and materials choices was an important aspect to the L462 project to keep pace with developments

L462 abandoned the split rear tailgate in favour of a huge sheet-moulded plastic tailgate, which allows a much more curved rear end. The offset licence plate was retained.

Land Rover Design – 70 years of success

in the premium SUV market. In addition to 18 body colour choices, two contrast roof colours were now developed, Santorini Black and Corris Grey, allowing added scope for personalisation.

Amy Frascella's team was also tasked with proposing five interior colourways, and decided on a theme where lighter colours were reserved for seat faces, upper surfaces of the IP and door top rolls in order to make it a more family-friendly interior where muddy marks and scuffs would not be so evident.

Frascella's team also developed an optional Dynamic Design Pack, with features including a contrast black or grey roof, more sporting front and rear bumper designs, luxurious Windsor leather upholstery and a sports steering wheel. An exclusive First Edition model of 2,400 was also offered at £68,295, available in three exterior colours: Namib Orange, Silicon Silver or Farallon Black. All came with a black grille and detailing, Narvik Black contrast roof and 21in wheels. Bespoke details included aluminium interior trim on the facia and doors featuring

L462 Discovery interior. "It has a strong vertical overlapping beam, we put the switchgear into primary zones, the rest of the cabin is about giving customers maximum versatility to use it as they see fit," says Mark Butler.

A fold-down inner tailgate and small underfloor locker were new features. The powered seat folding could be controlled by buttons located in the side of the trunk, at the C-pillar, via the infotainment touchscreen, or remotely using a smartphone app. The InControl remote key function was another alternative for remote car controls.

an etched map motif in celebration of New Discovery's British design and engineering credentials.

The engine line-up was based around the latest four-cylinder and six-cylinder engines, with vastly improved fuel economy over Discovery 4. The basic diesel engine was the 180ps 2.0-litre Ingenium Td4 engine, capable of delivering a combined fuel economy of 47.1mpg. A more powerful twin-turbo Sd4 Ingenium diesel produced 240ps, while the six-cylinder Td6 diesel had 258ps. A single petrol engine was offered, the Si6, a 340ps 3.0-litre supercharged V6 as used in the Range Rovers.

When launched at the Paris Motor show in 2016, Gerry McGovern described the design: "New Discovery's flawless volume and proportions, sophisticated surfaces and precise detailing beautifully combine with engineering integrity to create a premium SUV that will resonate with today's customers."

The end of Defender production 2012-2016

In August 2011 the Defender was given yet another engine change in a bid to keep the vehicle saleable for a few more years. A new Ford 2.2-litre EU5 diesel engine replaced the EU4 2.4-litre diesel, bringing improved levels of performance refinement to the Defender. Despite the smaller capacity and reduced emissions, the new engine produced the same 122ps power as the outgoing 2.4-litre engine, making this the cleanest Defender yet produced. For traditional

Rapid expansion of design activities

The Discovery was launched at the Paris Show in September 2016. Officially, it is now referred to as simply 'New Discovery,' not Discovery 5.

Land Rover enthusiasts, the Soft Top was reintroduced as a standard model, too, available with a khaki or black canvas tilt.

The Defender's specification was further improved with new alloy wheels, ventilated disc brakes on all variants, and high-backed front seats, designed to improve back support over rough terrain. In addition to the 'S' and 'SE' trim levels, five new option packs were introduced to the Defender for 2012 to offer more bespoke combinations, including part-leather seats and a leather steering wheel across the model range. The five option packs comprised the following:

- Leather Pack – Part-leather seats and leather steering wheel available on all body styles for the first time.
- Interior Pack – Carpeted floors, cubby box, part-leather seats and leather steering wheel.
- Comfort Pack – Air-conditioning, CD player with auxiliary input and convenience pack.
- Exterior Pack – Brunel grille and headlamp surrounds, body-coloured roof, wheel arches and side runners.
- Off-Road Pack – ABS, heavy-duty rim and MT/R tyre; tow ball and under-ride protection bar.

The Paul Smith Defender 90 was created in March 2015. "I wanted deep rich colours, but at the same time I wanted them to work together, yet be surprising," he says.

Land Rover Design – 70 years of success

In March 2015 SVO collaborated with British designer and Land Rover enthusiast Sir Paul Smith to create a one-off Defender 90. Exterior panels were painted in 27 different colours and inspiration was drawn on everything from the British countryside to Defenders used by the Armed Forces.

In July 2015, Land Rover confirmed the ending of Defender production to coincide with the production of two million examples. Three new limited edition models, each celebrating a different element of the Defender's unique history were announced. The Defender Celebration Series comprised Heritage, Adventure and Autobiography Editions.

The Heritage Edition was offered at £30,900 in Grasmere Green paintwork with a contrasting white roof and a classic grille, Almond Resolve cloth upholstery with black vinyl sides, and HUE 166 graphics recalling the registration plate of the first ever pre-production Land Rover – nicknamed 'Huey.'

The Adventure Edition was available at £38,400. It was fitted with additional underbody protection, Goodyear MT/R tyres, aluminum and black detailing and a leather-trimmed cabin. Three colours were offered: Corris Grey, Yulong White or Phoenix Orange.

Finally, the two-tone Autobiography received a power upgrade from 122 to 150ps. Available only with the 90 Station Wagon body,

The 2015 Defender Celebration Series offered three limited editions, termed Autobiography, Heritage, and Adventure.

The final Heritage Edition was finished in a newly reformulated version of the original light green, now called Grasmere Green; seen here on Llandonna Beach with 'Huey.'

Grasmere Green inserts were included on the centre console. Seats were trimmed in Almond Resolve cloth upholstery with black vinyl sides and 'HUE 166' labels.

The unique two millionth Defender was produced in December 2015 and auctioned for £400,000. Note the 'Red Wharf Bay' seat graphics.

it had full Windsor leather upholstery, Goodyear MT/R tyres, neutral tail and indicator LED lamps, a shiny metal trim package and more serious underbody protection. At £61,500 it was the highest list price ever charged for a Defender in the UK.

The actual two millionth Defender was produced in December 2015 and auctioned for £400,000. One month later, the final Defender – a Soft Top 90 – rolled off the Solihull production line at 9:22 on Friday 29 January 2016.

Land Rover Design – 70 years of success

DC100

One month after the launch of the 2012 model year Defenders, Land Rover showed the pair of DC100 concepts at Frankfurt Motor Show. The design team was headed by Richard Woolley, with Mark Butler and Oliver le Grice as lead designers on the project. "DC100 aimed to send a signal to the world that Defender wasn't completely off our radar, we were thinking about it," says Woolley. "It was an exercise in exploring 'what if': what architecture it might have, what powertrain, or derivatives in future. Putting it back in people's consciousness, showing that Defender is alive and well in Land Rover."

DC100 stood for Defender Concept 100in, meaning the vehicles used a wheelbase exactly midway between the existing Defender models, and identical with the old Challenger project of 20 years previously. DC100 was the more conventional of the two, being a three-door Hard Top in silver with a white roof, while DC100 Sport was a two-seat open fun buggy, painted in yellow.

The two show cars were built at prototype shops in Coventry, the Hard Top at HPL, the Sport at Concept Group International (CGI) – the same supplier who had built the LRX show car. In total, three or four models were built to show the diversity of Defender, and at least two were running prototypes, fitted with prototype 2.0-litre Ingenium engines that could be driven for demonstrations at slow speed. DC100 was billed as being a diesel and DC100 Sport boasted a petrol engine.

The interior used the same chunky, chamfered forms as seen on the exterior. The IP was a functional non-handed beam, with the gearshift mounted high up off the floor in a central console, flanked by a pair of orange grab handles. Seats were recessed into the rear bulkhead. The floor and rear loadspace were lined with a mat, pierced with rivets to provide some rugged protection. A metal toolbox was fitted in the rear.

The open Sport version followed the same design themes, but introduced some alternative bumper graphics and materials in the interior to project a more fun aspect to the design. A large lift-up rear deck covered the loadspace and could be removed if required.

Over the following months, the DC100 project was shown in various formats at different motor shows, including Delhi and New York. The Hard Top was later repainted red, while a van and explorer versions were also shown to demonstrate the diversity of the design.

DC100 was shown at Frankfurt Motor Show in September 2011. 'Cool and Tool' was the tagline. 'Cool' was the yellow DC100 Sport, while 'Tool' referred to the silver Hard Top. They were built on shortened Range Rover Sport T5 chassis.

DC 100 design team. (Left to right) Richard Woolley, Oliver le Grice, Gerry McGovern, Phil Higgs, Joanna Keatley, Mark Butler.

Rapid expansion of design activities

Hard top interior was more sober than the Sport, but still included bright orange inserts for doors and grab handles. The interior design was led by Mark Butler, with Joanna Keatley for colour and trend and Phil Higgs for interface design.

Sketch of DC 100 and DC100 Sport,

In March 2018 one of the DC100 models was destroyed in a fire at the warehouse in which they were stored.

Chapter

9

Current Land Rover design

Just as the public was thinking that Range Rover had done well to expand the portfolio to three models, another new variant made its debut. In a 2015 interview with *Director* magazine, Gerry McGovern said "By 2020 there's going to be 22 million SUV-type vehicles sold globally. So that's a massive market. We're asking: what are the products we could be creating that don't actually exist yet – like the Evoque, which we didn't have in our portfolio before."

The result was the Range Rover Velar, which premiered on 1 March 2017. This model was another example of a 'white space' vehicle – one that fills a previously unidentified niche, just like Evoque. The Velar put a new emphasis on glamour, modernity and elegance for the brand, a deliberately more urban target than the large Range Rovers. The launch was just one week before the Geneva Motor Show, at London's Design Museum, where it formed part of an exhibition entitled – appropriately – 'Reductionism.' "It's a transformation from the jungle to the urban jungle," said McGovern at the launch.

Rather than use the full-size D7u architecture, the Velar was designed around the smaller JLR D7a, or iQ[Al], platform that underpinned the Jaguar F-Pace, XF, and XE models. This was another all-aluminium platform, but used slightly narrower tracks and a 2874mm (113.1in) wheelbase, some 48mm less than the large Range Rovers. The basic vehicle underpinnings were similar to the F-Pace, and the two vehicles were designed from the start to share components and to be built on the same line at Lode Lane. Like the Jaguar models, the Range Rover Velar was engineered to use JLR's Ingenium line of 2.0-litre four-cylinder diesel and petrol engines, plus JLR's 3.0-litre six-cylinder engines.

Why a fourth model for the Range Rover line-up? First, there was a yawning gap in typical transaction prices between an Evoque at £45,000 and the Range Rover Sport at £70,000, something that Porsche was filling very successfully with the Macan. Second, the availability of the Jaguar F-Pace project (codename X761) meant that much of this smaller SUV hardware, engines and powertrain development could be easily shared, meaning it was an obvious model for Land Rover to develop fairly quickly at a reasonable development cost.

If the business case for it was compelling, defining the exact design direction for this project, codenamed L560, would need time to get it spot on. Work began in 2013, in parallel with the F-Pace. However, due to Jaguar's urgent need for an SUV model in its line-up, the F-pace project was given priority for production, allowing the Land Rover team a little more time to finesse the Velar design.

The exterior was a development of the Range Rover Sport theme, but lower and sleeker. More emphasis was placed on flushness of surfaces and fewer offsets between surfaces, especially in the front bumper and around the upper doors and greenhouse. Here, the focus was on precision shutgaps and demonstrating a high level of technology to create a new design purity. "We wanted to elevate Range Rover's design DNA to a new level," said Massimo Frascella, now Land Rover's Exterior Creative Director. "We were

Current Land Rover design

Range Rover Velar sketch. Named after the 1969 Range Rover prototypes with their fake identity badges, the Velar name derives from the Latin verb 'velare,' meaning 'to veil' or hide. The official pronunciation is 'Vell-Ar.'

Velar sketch builds on family themes, but in a far more minimalist fashion. The Velar was 50mm shorter and a full 115mm lower than Range Rover Sport – the same height as the Evoque.

Land Rover Design – 70 years of success

The continuous shoulder that runs around the rear of the vehicle very much defines the character of the Velar. Sitting the greenhouse inboard reduces its visual mass.

looking for a new level of simplicity, with the flush door handles and slim LED lights. We wanted an emphasis on the classic Range Rover proportions. The short front overhang and long tail give an incredible elegance, moving the emphasis rearward like luxury yachts, which make it sophisticated, strong and so pure."

To emphasise the long and low look to the vehicle, the upper greenhouse was designed to be as flush as possible, with gloss black door cappings and mouldings used throughout. In addition, the Range Rover floating roof design could be specified in either body colour or a Narvik Black contrast finish. With either colour, a fixed or sliding panoramic roof could be ordered that highlighted the Velar's airy interior space by allowing natural light to flood in. The fast windscreen rake and flush door handles helped the Velar to achieve a low 0.32Cd for the base model Ingenium D180 – the best drag figure in Land Rover's history.

The interior team was led by Mark Butler, and this time around the design process shifted somewhat to focus on finding new ways to demonstrate the latest technology and display screens that would be fitted to the vehicle. A 'speedfoam' model was produced to allow the team to explore the volumes of the interior space. "We started to look with the research team at the interior architecture and what we could do with the screen technology. There was a huge push to take a real step forward in the technology presentation of the vehicle," said Butler.

The overall aim was to further reduce the visual clutter of the interior, using the familiar Range Rover unbroken beam across the IP, but this time using very slim air vents and smooth, unbroken surfaces wherever possible. Touchscreens, head up displays and capacitive touch-controls on the steering wheel would help to reduce the visible switchgear to the minimum.

Current Land Rover design

A sketch showing an early proposal for the central screen, with a broad unbroken beam and a slim gap containing the air vents.

Final design. The Touch Pro Duo comprises two stacked 10in touchscreens. The upper screen deploys out of the upper IP beam when the vehicle is fired up. The lower screen is integrated into the centre stack with glossy curved edges and a pair of 'magic ring' controllers, inspired by high end camera lenses.

This sketch brings the console touchscreen closer to the driver, but is not so neat.

The central screen, head up display and capacitive touch-controls on the steering wheel all help to reduce the visual clutter.

After about four months a second demonstration model was milled with functioning screens, but without all the touch features. This was so the interior concept could be effectively communicated to others within the company, such as Product Planning or Marketing, who would need to understand the precise aims of the designers and experience the interior ambience for themselves. At this stage, the upper and lower touchscreens were still integrated as one piece. "That brought all the lower volumes a bit close so the cabin felt a bit claustrophobic," says Butler. "Everybody 'got' the technology, but we wanted to open up the space a bit more."

At this point in mid-2014 two decisions were taken. The first was to split the screens and have the upper screen flush with the IP beam, but able to deploy within comfortable reach once the systems were fired up. The second decision was to employ Panasonic's latest 'Magic Rings' rotary control wheels that appear to float on the touchscreen surface. This allowed the lower console screen to be moved forwards to free more space. The rings normally operate the climate control and cabin temperature is displayed in the centre, but when other vehicle functions are selected, one ring can be used to vary the Terrain Response setting, for instance.

Land Rover Design – 70 years of success

The interior team had been seeking alternative seating materials to leather for some time, and the Velar provided the right opportunity to move ahead on this. Amy Frascella explains the thinking: "The definition of luxury materials is changing, and what customers value in the products they buy is changing as well. We had to be ready for that."

Thus, Frascella's team collaborated with Danish textile company Kvadrat to develop a new high end seat fabric, a blend of wool and polyester that was soft yet durable. Available as Premium Textile option, it was offered as a £620 extra over the standard leather trim – a complete turnaround from earlier years, when fabric was the cheaper option. A second suedecloth fabric was also offered, manufactured using recycled plastic bottles, reflecting the shifting tastes away from animal products. Wood veneers still featured in the Velar interior, but were updated. New finishes included a pale open-pored veneer, a silver-streaked high-gloss grain, and a carbon fibre finish with interwoven copper filaments.

As with the Discovery, a 'First Edition' was offered that was available worldwide for one year only at around £85,000. Even more luxurious than the HSE specification, this model had the choice of 3.0-litre V6 petrol or diesel engines, and featured a wealth of extra features as standard, including full extended leather interior trim to complement the perforated Windsor leather seats in Light Oyster or Ebony, 1600W Meridian Signature Sound System, copper-coloured body trim accents, matrix-laser LED headlights and 22in split-spoke wheels. Three exterior paint colours were available for the First Edition: Corris Grey, Silicon Silver, plus a Flux Silver satin paint finish.

2018 Range Rover and PHEV

For 2018, the Range Rover was face-lifted with a revised front end, featuring a new grille with gloss black surround and Atlas-style mesh, flanked by the latest pixel-laser LED headlights. The front bumper was redesigned with wider lower vent blades and

Massimo Frascella, Director of Exterior Design.

Interior of Velar, with Kvadrat fabric and Oyster suedecloth. The Velar was awarded Production Car Design of the Year by *Car Design News*. This award is granted by a panel of judges who are all leading car designers, rather than journalists.

Current Land Rover design

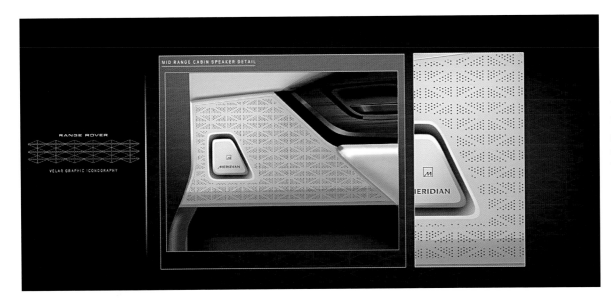

This close-up of the Meridian speaker illustrates the level of detailing that goes into the design process these days. The diamond graphic used throughout the interior can also be interpreted as a Union flag motif.

Gerry McGovern: Modernism and architecture

After returning to the UK from Los Angeles, Gerry McGovern decided he would fulfil a long-held ambition and design a house. The three-year project was carried out with help from his architect friend Adrian Baynes, and was inspired by the work of Californian architects such as Pierre Koenig's Stahl House, and the Palm Springs Kauffmann House by Richard Neutra.

The neomodernist house is sunk into a steep west-facing site not far from Gaydon and the Burton Dassett hills. The original dwelling was gutted and totally remodelled to provide a house full of space and light and is furnished with iconic pieces of mid-C20th furniture. These include Mies van der Rohe Barcelona and Harry Bertoia 'Bird' chairs, Florence Knoll chairs from the mid-'50s and Eero Saarinen's 'Tulip' chair.

McGovern starting buying modernist furniture and art in the 1990s. He has Jonathan Adler ceramics and 1950s Italian glass, while on the walls are prints by Patrick Heron and Patrick Caulfield. His Eames lounge chairs inspired his Lincoln interiors. "This place represents my design philosophy," he explains.

"You know I'm a modernist, and I do have a philosophical view when it comes to car design, that it is about volume and proportions. It is about reduction," he explained to *Car Design News*. "Every line should be there for a reason. There should be visual logic. I'm not a great advocate of what I'd call the Zorro school of design where there will be a slash here and a slash there, trying to be different for the sake of it. Land Rover has never been that way."

Talking to *Autocar* editor Steve Cropley, he added "What moves Velar on is its modernity. Modernist doesn't mean contemporary in the design sense. Modernism is a movement, a philosophy involving the simple use of forms. It's about paring back, about not including things unless they have a job to do. Or several jobs."

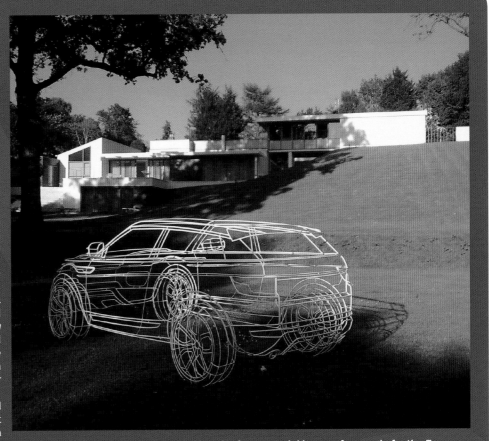

The house was completed in 2008. The wire frame model is one of 40 made for the Evoque launch.

Land Rover Design – 70 years of success

The Range Rover was face-lifted for 2018, with the latest pixel-laser LED headlight style. This shows the P400e PHEV version.

the bonnet was modified to be marginally longer to suit. New tail lamps completed the exterior revisions.

The interior was updated with wider seats and redesigned door linings for more storage space and fitted with the latest technology for greater comfort and convenience. This included Touch Pro Duo, combining two high-definition 10in touchscreens on the centre console, where information could be swiped from one screen to the other. Further new features were a gesture-controlled roof sunblind, air cabin ionisation and an Activity Key, with which customers could securely lock and unlock their vehicle without the need to carry a conventional key fob.

Executive Class Seating was available on Vogue SE models and standard in Autobiography derivatives. Also new for 2018 was a power-deployable centre console that could sink into the floor, allowing a fifth passenger to be accommodated.

Two plug-in hybrid electric vehicles (PHEV) were also offered: the Range Rover Sport and Range Rover P400e, available in both standard and long-wheelbase body styles. The P400e replaced

In February 2018 a Sport P400e PHEV climbed the 45-degree, 999-step staircase to Heaven's Gate rock arch in China to prove the exceptional capability of the drivetrain, with 404PS and 640Nm of torque.

the SDV6 Hybrid Diesel powertrain, and provided sustainable performance by combining a 300ps (221kW) 2.0-litre four-cylinder Ingenium petrol engine with a 116ps (85kW) electric motor. Thanks to its electrified powertrain, the Range Rover P400e emits only 64g/km on the NEDC combined cycle, and provides an all-electric range of up to 31 miles (51km) without the petrol engine running.

Special Vehicle Operations (SVO)

From the earliest days, Land Rover always had a bespoke operation to produce specialist versions. By its very nature the Land Rover lent itself to being adapted and customised to suit specialist needs, be they agricultural, military or luxury customers.

In 2014, a new JLR Special Operations division was set up to cover both Land Rover and Jaguar, with John Edwards as its Managing Director. Seasoned design manager Richard Woolley was delegated to head up the Land Rover design operation, based at a separate off-site studio in Coventry. "Special Operations encompasses four sub divisions," he explains. "Special Vehicle Operations (SVO), Classic Heritage business, vehicle personalisation, finally branded goods and merchandise. Anything that has design content across those I look after."

"SVO's remit is to amplify the qualities of the three pillars we have," he continues. "Gerry is keen to stress they're not the better versions of the cars, but rather they amplify certain attributes of the core products, namely the luxury, capability and performance." Thus, Luxury is handled under the SVAutobiography name for Range

Land Rover Design – 70 years of success

Rover, off-road capability comes under SVX, and SVR is about reduced weight, increased power and improved aerodynamics.

As mentioned in the previous chapter, the Range Rover Sport SVR was the first product to be developed in August 2014, with the 5.0-litre supercharged V8 uprated to 550ps. The Range Rover SVAutobiography was introduced the same month. In 2016 the 550ps V8 engine was installed in the Range Rover to create the Range Rover SVAutobiography Dynamic, a new flagship for the range, using the standard wheelbase body. Then in 2017 SVO showcased the Discovery SVX concept at Frankfurt Motor Show, with longer travel suspension, chunky all-terrain 20in Goodyear Wrangler tyres, an integrated power winch and roof bars.

Range Rover Sport SVR is currently the division's most successful product, selling over 2500 units a year, while the Range Rover SVAutobiography sells around 1500 models per year, all at around £100,000 apiece. £40m of additional revenue is no small beer, and dwarfs Spencer Wilks's original Land Rover SVO dreams of 200 extra vehicles per year back in 1957.

SVO models are built at the new SV Technical Centre outside Coventry, otherwise known as Oxford Road, formerly the site of the old Rootes factory. It has its own paint facility, a £20m investment that allows a much wider choice of special paint colours and duotone paint finishes. "There is also re-trimming of interiors there, too, with dedicated craftsmen. They can respond more quickly now. The old SVO facility in Solihull was bursting at the seams, customers kept coming and asking how they might spend more money on their Range Rover, yet we only had a limited capacity. We could see the opportunities with aftermarket converters such as Kahn or Overfinch, now we can do all that in-house and deliver it with same level of quality and integrity as the core cars," says Woolley.

The current Land Rover SVO design team is 50-60 strong, including 15 creative designers, CAS modellers, clay modellers, studio engineers and programme management. This demonstrates the importance of bespoke design and manufacturing to Land Rover's current business: this specialist studio is comparable with Land Rover's entire design operation from the 38A studio a couple of decades before. "We have ambitious plans for accessories, more line fit so the vehicle turns up at the dealer with much of that already built in. We can now do much of that on the production line," explains Woolley.

Discovery SVX concept was shown in 2017 to showcase 'Premium Durability.' New features include an integrated winch and roof bars and 20in forged alloy wheels. The show car was finished in Tectonic Grey satin paint finish with Rush Orange accents.

The 2018 Range Rover SVAutobiography Dynamic is the most powerful production Range Rover to date. Power output was increased to 565ps from its V8 supercharged engine, with 700Nm of torque available. That's enough to deliver 0-60mph in just 5.1 seconds. The Executive Class seating configuration is finished in exclusive quilted perforated semi-aniline leather trim available in four sporty colourways: Ebony, Ebony/Pimento, Ebony/Cirrus and Ebony/Vintage Tan. It also debuts two new veneer choices, Steel Weave carbon fibre and Grand Black, with finishers in dark brushed aluminium also available.

In March 2018, Michael van der Sande replaced John Edwards as the new Managing Director of the Special Operations division, which now has a total of 500 staff across all four sub divisions. He brings with him experience from Aston Martin, Harley-Davidson and Tesla. He most recently oversaw the development of the new Alpine A110 sports car, and thus brings a clear idea of how a bespoke division will need to adapt in a changing world of EV architectures and shifting expectations for personal mobility.

Current Land Rover design

Michael van der Sande replaced John Edwards as the new Managing Director of Special Operations division in March 2018.

Range Rover SV Coupé

At the Geneva Motor show in March 2018, Land Rover unveiled the latest version of the Range Rover, designed by Woolley's SVO team. The SV Coupé is a celebration of the Range Rover bloodline, with a striking two-door silhouette which alludes to its heritage – the original Range Rover from 1970 – while being thoroughly modern and contemporary.

"It's always been the ambition to do our own vehicle. In the lead up to the 50th year of Range Rover we felt it would be great to do a very special edition, so we came up with idea of recreating the two-door," explains Woolley. "It was very quickly decided that it would be a limited edition of 999, not a full production vehicle, and would be hand built at Oxford Road."

One of the original starting points was to see whether the Range Rover Sport screen and roof could be mated to a full Range Rover lower body to achieve the sleek coupé look. In the end, all the exterior panels are new, with only the bonnet and lower half of the tailgate unchanged. The vehicle is 8mm lower and 13mm longer at the rear compared to a standard four-door Range Rover. It is also offered with a couple of firsts for Land Rover: 23in wheels and power-closing doors. "With such long doors [1.4m], you can't reach," says Woolley. "It's an essential."

The BIW is constructed and modified in Solihull, then transported to Oxford Road to be painted and assembled. It then goes down the normal production line. The running gear is basically standard but modified to suit the driving characteristics of the SV Coupé in terms of suspension tuning. Woolley admits there were many subtle changes. "We were able to manipulate the design with a faster screen angle, a slightly higher belt line. We worked closely with the engineering team in SVO. It's the integration of design and engineering that makes the brand special."

Together with a supplier, SVO has also created a new veneer inspired by boat design. Known as Nautica veneer, it will eventually be rolled out to other special models. This uses a special patented form-following process to fuse together walnut and sycamore wood. "Because we knew we were making just 999 we could use special tooling methods on the SV Coupé. For example, exterior metal components are machined from solid, then hand polished," explains Woolley.

"We've done previews in London and Paris, also Qatar and Dubai. It is very much a Collectors car. The clientele typically already own a Range Rover, together with a collection of other cars, too. They're brand fans, they love Range Rover. With only a few available they really want one."

The SV Coupé has many differences from the standard Range Rover, including faster front and rear screen angles. Only the bonnet panel and lower rear tailgate are carried over.

Land Rover Design – 70 years of success

The SV Coupé has a unique front bumper with a deeper front grille surround. It also has slightly more tumblehome in the side glass than the four-door version. SV Coupé will be hand-assembled at the SV Technical Centre at Oxford Road, Coventry, with just 999 examples being produced, with first deliveries due in autumn 2018.

Detail of the side vent, which is moved to the front fender on the SV Coupé. (Author's collection)

The door panel features Nautica veneer, a combination of walnut and sycamore wood. Just visible is the deployable step that powers out from under the sill to aid entry. (Author's collection)

Current Land Rover design

The SV Coupé interior. The front seats are in lighter colours with dark colours in the rear. "It's the opposite of a chauffeur-driven arrangement. It's amplifying that sense of peerlessness," says McGovern.

A new Defender ...

There have been several attempts to replace the Defender, going back as far as the SD5 in the 1970s. Ibex/Inca, Challenger, Heartland and LCV 2/3 all proved false starts and were abandoned along the way, not to mention the promising DC100 concept. The stumbling block has always been the business case to support the huge investment required for a unique chassis and architecture on which to develop a range of vehicles in a variety of body formats and wheelbases, which can meet the exceptional capability expected of a Defender. Production of the vehicle had settled at less than 20,000 units per year, way too low to justify such a major investment. "That's why it's taken so long to figure out the business case to do it, wanting do the right thing for the brand and for the business, a tricky balance," admits Richard Woolley.

The reaction to the DC100 design in 2011 was mixed. Dyed-in-the-wool Land Rover enthusiasts saw it as too superficial and toy-like, while others felt it was a parody of the simplistic form language of a Defender, lacking sufficient design values of its own for the 21st century. "It started the realisation that whatever we do with Defender it will be a different vehicle, the world has changed from 1948, the vehicle needs to be a different proposition," he continues.

Land Rover acknowledge that a new Defender is waiting in the wings, and Woolley provides hope that it will remain true to the brand. "We acknowledge our burden of responsibility to keep that as part of our portfolio. Defender owners have very fixed opinions about what Defender should be. And – much as we love them – they are a dwindling bunch, with sales down to 17,000 from a peak of 40,000-plus in the late 1980s. Lots of people love it, but only a small number actually bought one and use it on a daily basis. We need to replace it with something that not just regains that level, but brings new customers into the brand. We need to make it a relevant vehicle that meets people's expectations, reflecting the way the market has shifted. And of course all the other current concerns: pedestrian safety, emissions, lightweight architecture etc. So a clean sheet of paper is needed. It's a modern interpretation, not a pastiche. None of it is literally the same. Hopefully, with new Defender we can bring the diehards along with us, we need that kind of enthusiasm."

When talking to American Land Rover enthusiast magazine *Alloy+Grit*, Gerry McGovern said "I think what the new Defender has to do is maybe replicate the charm and the essence of the original, and that's more around its capabilities, its durability, its ruggedness. And we have to be careful we're not preoccupied with trying to do a retrospective vehicle or trying to do a modern-day facsimile of

223

Land Rover Design – 70 years of success

DC100 was later shown as a full expedition vehicle, with air intake snorkel and integrated roof rack. Land Rover acknowledges that a new Defender is waiting in the wings. "We acknowledge our burden of responsibility to keep that as part of our portfolio," says Richard Woolley.

what's gone before. So we shouldn't be doing the Mini thing, or some of the other [retro] things we've seen.

"I think people will be quite surprised when they see it. I think a lot of people, even the traditionalists, will smile. But it has to be modern, it has to be of its time, and it has to be able to do what it says it can do on its tin."

In the meantime, as part of the 70th anniversary of Land Rover, Land Rover Classic announced the Defender Works V8 in January 2018. These will be nearly-new vehicles starting with 2012-2015 model year donors, rebuilt to order featuring a V8 engine – the supercharged 5.0-litre as fitted to the current Range Rover, but retuned with 405ps.

Classic will hand-build up to 150 examples of this uprated Defender at the Oxford Road facility, complete with stiffer springs and anti-roll bars, plus upgraded brakes to deal with the extra performance. Sawtooth 18in wheels wearing 265/65 all-terrain rubber will be fitted. Design upgrades include bi-LED headlamps, machined aluminium door handles, bonnet lettering and fuel filler cap. Eight standard colours are available, but all will be finished with Santorini Black for the roof, grille and wheelarch flares.

The changing role of the Design Department

The prevailing attitude from the 1950s up to the 1990s was that the engineer's job was to create products that customers would appreciate for their functional advances in technology, packaging and features. Styling was a necessary but subordinate evil. The engineers regarded themselves as 'the designers.' Alec Issigonis as chief engineer at BMC epitomised this approach, scorning not only the need for professional styling input but also denying that the needs and desires of customers should be of any concern. "A designer has only to make a good car that satisfies him, and if he is a practical man it will satisfy the world. I have never had any ambition in my life except to satisfy myself and never think of a new car in terms of the people who are going to buy it," he once said.

David Bache was one of the first 'stylists' in the UK to emerge from this stranglehold, and was fortunate in that the management

Works Defender V8, with 18in sawtooth alloy wheels. It is the first V8-powered Land Rover since the 50th Anniversary models. Defender 90 or 110 Station Wagon models can be converted, with prices from £150,000.

The Works Defender V8 interior features a redesigned centre console to accommodate the eight-speed ZF automatic transmission. Windsor leather is fitted to the Recaro sport seats, IP, door panels and headliner. Door panels date back to Dave Evans' redesign from 1983 – one of the longest-running interior trim items ever produced.

Land Rover Design – 70 years of success

Amy Frascella is head of the Colour and Trend team at Land Rover. "The definition of luxury materials is changing, and what customers value in the products they buy is changing as well. We have to be ready for that."

at Rover had a more positive ethos than other British car companies. Rover was not as autocratic as BMC, Jaguar or Triumph, nor as hierarchical as Ford or Vauxhall. The fact that he was well-educated and held views about how design could improve safety and ergonomics meant Bache became known to the media in the 1960s as someone who might be interviewed and could talk eloquently about design. However, he was not a household name in the way that Gerry McGovern or Ian Callum have since become, and was certainly not the celebrity that would be wheeled out to launch the car itself, as happens today. That was still very much the preserve of the Wilks brothers and the engineering directors, such as Spen King.

The role of the Land Rover Design Department has changed enormously in more recent years, too. Whereas before it was chiefly a styling studio tasked with producing sketches, models and ideas to support the design of new vehicle projects, today the remit is more about being the main creative visionary group for the entire company. Land Rover Design is a well-oiled machine these days, with highly developed processes and methods that enable it to develop projects at a staggering pace. It is a far cry from the old days when the Stage 2 face-lift to produce the One Ten Land Rover took seven years …

The composition of the Design team has remained remarkably stable over that period. Most of the senior team in Design started with the company over 30 years ago, and possess an unrivalled wealth of experience and understanding about the product and the brand. They are supremely qualified to reflect on how Land Rover design has developed, where it has come from, the struggles and the successes, and – most importantly – where it is headed.

So, what have been the big changes in the design process in the last 20 years? Phil Simmons is now Studio Director, Exterior Design Realisation, but is also responsible for design communications – an important new function so the design team can explain its ideas. "I guess we can point to what has not changed: full-size clay modelling, which remains the focal point of our development process is still the best material we know to make those hands-on interactive changes to the surfaces of a car. We've still got very skilled modellers making those changes by hand. However, the whole process is being speeded up by application of new technologies that allow us to both bring in engineering data and changes to models, and we can mill surfaces directly onto models in a fraction of the time it once took."

The studio has now moved to on-plate milling, which is becoming the industry standard method. "Not noisy milling in another shed, but whisper-quiet milling right here in the studio meaning turnaround time is vastly reduced," he continues "Eventually it means we can reduce the time taken to develop the vehicle. We can use the benefit in number of different ways: either do more models in the time allowed, or the same number in a reduced time. It depends on the programme. Sometimes you need time to mature an innovative or novel idea, so use the technology to enable you to do multiple iterative loops to get to a point where its 100 per cent feasible. We would not want to go back to the old ways, far too much manual labour involved!"

By using virtual tools for design and simulation, there has also been a dramatic reduction in the need for physical prototypes. Whereas in the past several hundred hand-built prototypes would be made, today much of the design and development work can be carried out digitally, or using simulators, leaving the physical testing work to be done on a smaller number of prototypes built off low volume tooling, often using 3D printed components.

Like Simmons, Alan Sheppard is another member of the senior team who has returned to the company from outside and is able to reflect on how it has changed over the last 20 years. After his involvement with L322 Range Rover, he stayed on with BMW and became a manager with the Rolls-Royce design team that was being established in Munich. He returned in 2017 and is now Studio Director, Interior Design, Design Research and Innovation.

"Coming back to Gaydon, what I find is I'm meeting more intelligent, clever people with triple the amount of mental agility and invention, great problem-solving skills, absolutely amazing. The amount of talent here is incredible, the company is three times the size of when I left, the factories are state-of-the-art, the facilities are great."

Dave Saddington reflects on how the Gaydon studio has adapted and changed: "Canley had little security, Gaydon was so much better. Our own viewing garden, huge windows. It has stood the test of time. The only trouble now is that with more screen-based design activity, there's too much light! This studio with its workshops next door and viewing garden where you can stand 250 metres away for viewings … You can drive here onto the test track and back into here again – it's

Current Land Rover design

Paint colours are initially explored using 3D form models, and the permutations have grown with the use of contrast roof colours and lower side mouldings. A paint colour takes around three years to develop from first trend input, through feasibility, weathering and then mastering before it hits the road.

great. We know neither BMW, Ford nor Tata have anything like this in all their sites. Quite a jewel in the crown here, we're very proud of it."

Attitudes towards design have changed, too. Andy Wheel explains: "The derision towards designers, the names – that doesn't happen around here anymore because so much of the business realises there are many people in this department who have got the scars from previous programmes, the business acumen as well as the creativity. That's a consequence of the hard times in the financial crisis."

Richard Woolley agrees: "I think because of the learning experience with Honda, BMW and Ford we were able to take the helm properly under Tata and bring the business to where we are today. Throughout all that turmoil of ownership, of management, of processes, the design department remained a core strength of the business right through all that."

"I remember Ravi Kant coming over and talking to the senior leadership in the Heritage Centre," adds Woolley. "It changed my whole perception of what was going on. These guys loved the two JLR brands, they told us 'you are the experts, just get on with it, deliver the products.' Not a dictatorial approach, but allowing us to become masters of our own destiny. When Mr Tata came to visit and invest his money with us it was a revelation and we haven't looked back since."

It wasn't always like that, and the arrival of Tata marked a fundamental shift. Being trained as an architect, Ratan Tata understood the role of design in other disciplines and asked the key question: 'Why does Design report to Engineering?'

To redress the situation, Gerry McGovern was appointed to the main board of JLR directors in June 2008, together with his opposite number, Ian Callum at Jaguar. Thus, for the first time, Design was represented at the highest level within the company rather than simply reporting through Engineering. "They are mutually compatible, design and engineering. That doesn't mean one having power over the other; it means having equality and working together," confirms McGovern. "We have a monthly meeting with the CEO, just the three of us, where we go through all the designs."

McGovern is keen to stress how Land Rover is a premium brand. Range Rover standing for luxury, Discovery for versatility, Defender for durability. "They're all premium," he adds. "By premium durability, Defender needs to be rugged. They're all incredibly capable but Defender needs to have that ruggedness in its visual sensibilities."

Gerry McGovern, David Saddington and interiors design manager Sean Johnson view an interior buck. Despite the use of digital tools, physical models are still highly valued as part of the evaluation process.

Land Rover Design – 70 years of success

Aerial view of Gaydon 2018. The site is currently undergoing massive redevelopment, with the relocation of many of Jaguar's engineers and designers from Whitley. GDEC block is on the centre left here. The car park in the centre is being filled in with a massive new R&D block and design studio for Jaguar. The circular building on the left is the British Motor Museum.

Looking ahead, what are the likely changes for design in the next ten years? Until a few months ago the likely answer would be that there will inevitably be more CAD used, with decisions on models made largely through CAD visualisations. Recently that view has been revised.

"We've shifted back to a slightly more traditional method, you get less surprises later on," says Massimo Frascella. "We've just gone back to producing quarter-scale clay models, then moving to full-size models. With younger designers, using scale models forces them to focus on the volumes, the proportions, rather than details."

The previous setup where the studio was split into separate creation and realisation design teams has also just been revised, with a single director now responsible for the whole exterior or interior design process, albeit with individual designers gravitating towards their core skills within the process. Frascella is now Director for Exterior Design: "That allows a more seamless transition, before there was a bit of a 'wall,' where one team would hand over to the other, saying 'now go and deliver it.' It worked, but there were inefficiencies."

Simmons gives the reasons why clay modelling is still valued: "There are examples in other industries such as architecture where design teams have moved to 100 per cent digital, from creation to final realisation as a product. We're still some way off going there. All the expertise we've built up means we focus on clay for very good reasons. All the tension and interaction in surfaces you just can't reproduce digitally. I'm happy to be proved wrong, but I'm happy to see clay models remain at the core of our design development workstream for many years to come, even if it gets more enabled by a digital input and milling machines to achieve the majority of 3D definition."

The current Land Rover Design department at Gaydon now totals 590 people, spread across all the disciplines, including creative designers for interior and exterior, Colour and Materials, HMI designers, clay modellers, CAS designers, SVO, studio engineers and support staff. In the main studio, the 190 creative designers sit on two levels, with mezzanine floors that have been constructed within the main studio to accommodate the increase in headcount.

Future research developments

The growth is not finished yet. Land Rover continues to explore 'white space' opportunities for new models to expand their portfolio in future. The Evoque is due to be replaced, and thoughts of another, more road-based flagship that might share architecture with the forthcoming Jaguar XJ have been mooted in the press, possibly reviving the 'Road Rover' name.

The greatest challenge will be the need to develop new architectures and EV powertrains as we move into the 2020s. JLR's ICE to ACE journey continues, moving from conventional internal combustion engines to Autonomous, Connected and Electrified vehicles. A number of initiatives are under development, and the new Jaguar I-Pace provides an indicator of the type of electric powertrain and architecture that might be deployed. Electric powertrains have a number of advantages for 4WD capability, including incredible torque at low speed, ease of packaging, simplicity and reliability.

Current Land Rover design

For Land Rover, the greatest benefit will be lighter weight, with the need to employ massive gearboxes and heavy transfer cases eliminated.

Electric powertrains also offer new benefits in many off-road situations such as working with animals, wildlife conservation and safaris, where the silence and stealth of the vehicle is a profound advantage. With climate change there is a growing need for vehicles that can cope with greater variations in weather and fast-changing road conditions, be it unexpected flooding, rutted roads, potholes or short-lived snowfalls in summer months. Land Rover is ideally placed as a brand: the desire for vehicles with higher ground clearance and more rugged capability across all sectors of the market is only going to increase.

By 2012, the advanced design team under Richard Woolley was renamed Strategic Design, and began to look at wider Land Rover branding issues, but after a couple of years its title was changed again to Design Research and Innovation, a JLR-wide group rather than just Land Rover, and based off-site at an anonymous industrial estate in Coventry. Here, the remit began to shift, with the introduction of team members from a non-automotive background to create an interdisciplinary team led by Oliver le Grice that would look at things from a wider perspective.

The idea was that they would bring another perspective to the debate on transport, the future of mobility and road infrastructure, thinking more on a business level than purely design or engineering solutions, looking beyond the current cycle plan into what is going to happen ten or even 20 years ahead. The team started with around 25 staff, including eight designers, technical researchers from the University of Warwick, eight visualisers/film makers, and digital modellers.

The reorganisation of design research followed the start of construction of the £150m National Automotive Innovation Centre (NAIC), which is scheduled to open at the University of Warwick in summer 2018, providing a state-of-the-art technology hub that will bring together JLR's current 1000-strong advanced research team and collaborative partners from the supply chain and academia. The 33,000sq m (350,000sq ft) Warwick facility is due to become the hub for JLR's advanced research with cutting-edge workshops, laboratories, virtual engineering suites and advanced engine facilities. In total, more than 8000 engineers, designers and technologists are now based at the two UK engineering and design centres at Gaydon and Whitley, together with the existing advanced research facility at the University of Warwick.

The National Automotive Innovation Centre (NAIC) is scheduled to open at the University of Warwick site in summer 2018. This will become the hub for JLR's advanced research with workshops, laboratories and virtual engineering suites.

Land Rover Design – 70 years of success

Alan Sheppard sums up the challenges ahead. "We've achieved the dream from the 1980s, becoming a successful independent company. It's about moving on now, a paradigm shift. What kind of an ecological system do we want to hand onto our grandchildren, how do we sustain our business, how do we deal with so much transportation on the roads? Those are fairly big issues, but also people's expectations are changing. What level of service do they require? What value for money? How much stimulation do they want from their vehicles? These are the challenges we've got to address."

From the earliest beginnings of Maurice and Spencer Wilks in 1947, David Bache's arrival in the 1950s, to the establishment of Tony Poole's independent styling studio in 1981, the story of Land Rover design is one of constant expansion and improvement in the way a vehicle is created from first sketches, through subsequent 3D models, and then to final launch. The changes in methods and processes have been immense, and continue to evolve as new digital tools allow more sophisticated virtual presentations to support physical models, and new ways of engaging the design story with customers and enthusiasts of the brand.

With Land Rover's creativity at an all-time high, and the resources to back it up, McGovern reflects on his achievements. "If you look at how we've evolved, we're on this journey of transformation. In a few more years this business will look nothing like it looked before. And Design has played a fundamental role in that. Why is that? The mindset is changing, the culture of going from what was a specialist brand to a brand that's more universally appealing. And one that uses Design as the conduit to achieve.

"At the same time, doing it in a way that maintains its essence and credibility is a tricky one. We have to keep the unique DNA that's evolved over time and present it in a far more compelling, more relevant way than it ever was before. I feel passionate about the importance of design and how it can enrich people's lives and, more importantly from a business perspective, how it can make businesses successful. If I blow my own trumpet it's because I've managed to make design in this company respected."

Gerry McGovern, Director of Design and Chief Creative Officer. His responsibility is not just for the design of new models – he has a wider remit of developing the look and feel of the Land Rover brand as a whole.

www.velocebooks.com / www.veloce.co.uk
Details of all current books • New book news • Special offers

Appendix 1

Land Rover model code numbers

The original Range Rover developed by NVC was initially known as the 'Alternative Station Wagon' project. It was later known as the '100in Station Wagon'. Pre-production prototypes in 1969 were registered and badged as 'Velar' in a bid to avoid the media guessing it was a Land Rover project.

During the 1980s Gilroy era, Land Rover projects were given bold, visionary codenames, reminiscent of military operations.

• Adventurer	1983 crossover project (abandoned)
• Capricorn	1983 forward facing rear seats in Ninety Station Wagon
• Aquila	1984 Range Rover EFi
• Llama	1985 Forward Control light truck (abandoned)
• Ibex/Inca	1985 Land Rover/Range Rover common structure (abandoned)
• Eagle	1986 Range Rover homologation for USA, including new front grille
• Discovery	Original codename for P38A Range Rover
• Pegasus	Revised codename for P38A Range Rover
• Jay	1989 Discovery 3D and 5D
• Challenger	1991 Defender replacement (abandoned)
• Pathfinder	1991 Freelander initial concept code – Land Rover version
• Oden	1991 Freelander initial concept code – Rover version
• Cyclone	1992 Honda Shuttle 4x4 prototype and interim name for Freelander project
• Wolf	1994 Defender XD (extra duty) for military use
• Romulus	1994 Discovery 1 face-lift
• Heartland	All-new Discovery 2 (abandoned)
• Tempest	1998 Discovery 2

In the late 1980s, two Land Rover projects were given location-based project codenames.

• P38A	1994 Range Rover Series 2
• CB40	1997 Freelander Series 1

After the BMW takeover in 1994, Rover cars were given 2-digit R-codenames eg R50 for new Mini. Similarly, Land Rover projects were given 2-digit L-codes, eg L30 was the all-new Range Rover.

- LCV 2/3 1999 Defender concept (abandoned)
- L30 2002 Range Rover
- L50 2003 Discovery 5-seater (abandoned)
- L51 2003 Discovery 7-seater (abandoned)

Land Rover Design – 70 years of success

After the Ford takeover, a series of three-digit L-codes was instigated, starting with L322 for Range Rover. This coding is still used today.

• L322	2002 Range Rover (renamed)
• L314	2004 Freelander Series 1 face-lift
• L316	2007 Defender upgrade
• L318	2002 Discovery 2 face-lift
• L319	2004 Discovery 3 (LR3 in USA)
• L320	2005 Range Rover Sport
• L359	2006 Freelander 2. (LR2 in USA)
• L420	2009 Range Rover Sport face-lift
• L538	2011 Evoque
• L405	2012 Range Rover 4
• DC100	2012 Defender concept
• L494	2013 Range Rover Sport 2
• L550	2014 Discovery Sport
• L462	2016 Discovery 5
• L560	2017 Velar
• L551	2019 Evoque
• L663	New Defender (Project Darwin)

Appendix 2

Glossary of design terms

While not an exhaustive glossary of automotive design terms, the following are terms used by designers and engineers and are found throughout this book:

- 4WD, also 4x4 Four-wheel drive.
- A-pillar, B-pillar, C-pillar, D-pillar The notation of the car's roof pillars, from front to rear. Whereas the A-pillar and C- or D-pillar are chiefly concerned with the style and flow of the greenhouse, the B-pillar is primarily functional, and thus often blacked-out to minimise its visual intrusion (see DLO).
- Alias The main CAD software used within design studios to create 3D digital models. The full title of the software is Autodesk Alias.
- Armature The frame or structure for a mock-up model or prototype. It is also used to mean the underlying substrate or structure for a component – for example the instrument panel.
- Belt line The main dividing line between the lower body and the greenhouse. Typically the base of the side windows (UK colloquial term = waistline).
- BIW 'Body in White' Originally, most steel bodies were painted in white primer at the initial paint stage. Nowadays 'BIW' refers to the bodyshell of the car either in a raw unpainted state or in its initial primer coat.
- Bulkhead The main structural firewall between the engine bay and the interior of the car.
- CAD Computer-Aided Design.
- Canson paper A type of coloured drawing paper, featuring a heavily textured surface. Popular in design studios in the '60s and '70s.
- Cant rail The structural section above the doors; the main box section at the edge of the roof.
- CAS Computer-Aided Styling. In some organisations it can also mean Creative Aesthetics and Styling – in other words the team that uses CAS tools.
- CKD Completely Knocked Down ie a complete kit of parts to assemble abroad as a vehicle.
- Conté crayon A hard drawing pastel made from compressed powdered graphite or charcoal mixed with a wax or clay base. They are square in cross-section.
- Dashboard See Instrument Panel.
- DLO 'Day Light Opening,' or the complete graphic shape of the side windows, including the B-pillar. Often outlined with a chrome finisher to emphasise the overall shape. This represents the main graphic element for the side view of the car.
- Door card Flat door inner trim panels, made of cardboard or hardboard until the 1970s. Now universally replaced by moulded items – more correctly termed door trim panels.
- Di-Noc A highly stretchy 3M plastic film used to cover clay models to represent a painted finish. The film is typically painted in silver, although other colours can be used, eg dark grey for windows and headlamps. Di-Noc is also used as a quick tool to check highlights as the clay model is being developed.

Land Rover Design – 70 years of success

- Escutcheon The surrounding to a handle or knob, such as a door armrest or interior door-release handle.
- Fender The front or rear panel surrounding the wheelarch, otherwise known as the wing. Due to the global nature of automotive design and the influence of the Pressed Steel Company (who always employed this term on drawings), the American word 'fender' is used in design studios.
- Foamcore A proprietary type of mounting board, typically used in graphic design studios. Available in various thicknesses, it can be easily cut with a sharp knife or scalpel.
- Forward Control The steering column and pedals are mounted forward of the font axle, with the driver usually seated directly above the axle.
- FWD Front-wheel drive.
- Greenhouse The glazed upper body of the car – in other words the entire cabin above the belt line, from the base of the windscreen to the base of the rear window.
- Graphics Any graphic shape or feature used on the surface of the car, such as the grille, the shut lines, the DLO, the windscreen or rear screen, the headlamps, or any louvres.
- GRP Glass-Reinforced Plastic, also called fiberglass (US) or glassfibre (UK).
- H-Point/Hip Point This refers to the seating reference point, which is the hip joint of a standard seating mannequin. A number of regulations concerning packaging and vision angle are derived from this important interior datum point.
- Hard points Fixed engineering points on the car, determined often as a result of structural or legislative constraints, typical examples being the placing of the front bumper beam and the design of the crush structure.
- Header The structural section above the windscreen or rear screen, the main box section at the ends of the roof.
- Instrument Panel The panel below the windscreen containing the instruments. The meaning has changed over time. Originally used to describe the (usually central) panel that contained only the instruments, it has now evolved to become the main term used to describe the entire dashboard assembly running across the whole width of the car. For consistency, we have used this term throughout the book, rather than 'dashboard.'
- IP Shorthand for Instrument Panel.
- Lightlines Paths of reflected light that run along a surface and make it possible to understand its sculptural form without reference to its outline shape.
- Lofting The drawing up and smoothing-out of body sections, full-size. Derives from shipbuilding parlance, where the ship's hull timbers would be drawn up in the roof area of the shipbuilder's workshop – the only area with enough clear floorspace to do this.
- LWB Long-wheelbase – in other words, an extended-wheelbase version of the standard car.
- Mock-up The initial development model for any new car design. A non-functional representation of the exterior or interior of the car. A mock-up can be constructed from many different materials including sheet metal, solid wood, clay, foam or glassfibre, depending on the eventual function and method of build available to the designer.
- Package The overall layout of the car, including the occupants, engine and wheelbase etc.
- PHEV Plug-in Hybrid Electric Vehicle
- RWD Rear-wheel drive.
- Scuttle Upper area of the bulkhead, directly below the windscreen. It contains the wipers, washer nozzles, bonnet hinges, air intakes for the heater, and so on – a complex functional area that requires careful design to integrate these features into a coherent whole. Also called 'cowl' (US).
- Shoulder line The intersection of the shoulder surface beneath the side windows and the vertical door surface. Not to be confused with 'belt line' – which (confusingly) lies above the shoulder.
- Shut line The gap between two panels. Also called 'shut gaps' (US).
- Six-light A six-light cabin has three windows on each side of the car – for example, the current Range Rover. A car with just two windows is a four-light style – an example being the Defender 90 Station Wagon. The third window itself is also referred to as the 'six-light.'
- Speedfoam A design model milled out of dense polystyrene foam for a quick evaluation of basic proportions and volumes of the design.
- Stance The way in which a car sits on its wheels – a combination of proportions, wheel-to-body relationship and wheel size.
- Styling buck A mock-up model of an exterior or interior, typically using clay. Interior bucks may be part-models, eg just the front cabin and front seats.
- Tape drawing A 'drawing,' usually on Mylar transparent film, using photographic black tape of different widths. The tape drawing allows the designer to precisely refine and correct the lines of the car as it is translated into a 3D clay model. The slightly flexible nature of the tapes means that gentle curves can be created, the radius depending on the width of tape selected.
- Tilt A canopy for a wagon, boat or stall. Usually made of canvas or PVC. Derives from Old English teld, a tent or canopy; akin to Old High German zelt, meaning tent.
- Trunk Luggage compartment or boot. Due to the global nature of automotive design and the influence of Pressed Steel (who always used this term on drawings), the American word 'trunk' is invariably employed these days in design studios.
- Tumblehome The slope of the side windows from vertical. Originally derived from naval architecture, where the sides of a galleon slope inwards above the water line.

Also from Veloce Publishing –

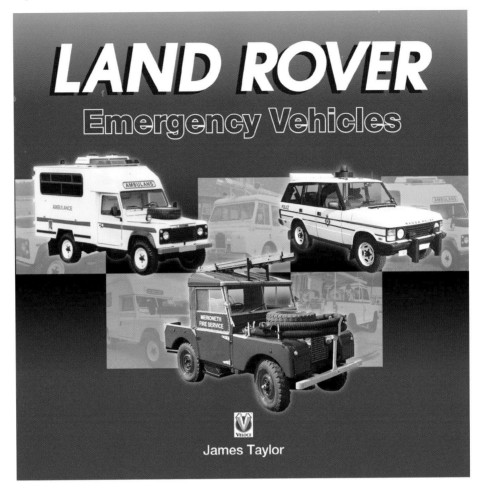

An historic and nostalgic look at the role of the Land Rover in emergency services over the last 70 years. Land Rover products have been used by the emergency services almost from the moment the first model left the factory in 1948. The agility and size of these vehicles made them an immediate hit with fire services, where they initially became popular as factory fire tenders. Police forces were also attracted by the cross-country ability and versatility of Land Rovers, especially outside Britain, and, when long-wheelbase models provided extra space, they also became favourites for ambulance conversions. This book will interest Land Rover enthusiasts and emergency-vehicle enthusiasts alike, with evocative photographs that illustrate both historic vehicles and more recent vehicles in action.

ISBN: 978-1-787112-44-5
Hardback • 25x25cm • 144 pages • 367 pictures

For more information and price details, visit our website at www.veloce.co.uk • email: info@veloce.co.uk • Tel: +44(0)1305 260068

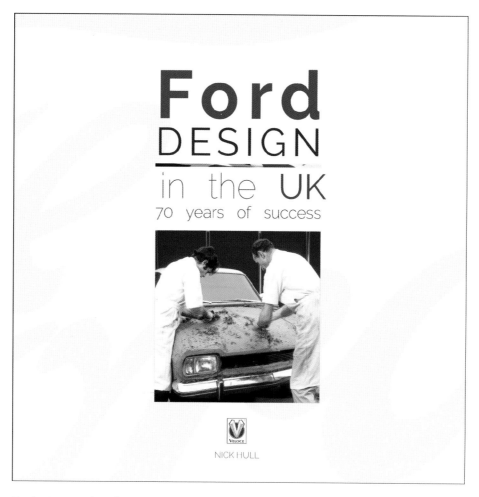

Exploring Ford UK's design studios during the past 70 years, this book provides a unique insight into the company's history, its UK studio locations, and delves into the lives of the designers, modellers and studio engineers.
As a profession, automotive design has changed hugely over the past century and continues to evolve as new processes are developed. This book charts the development of Ford projects in the UK, particularly those designed in the Dunton studio, which opened in 1967 and is still a key part of Ford's design resource in Europe. From the early days of chalk drawings and wooden models for the Consul to today's digital renderings and milled clays for the latest Transit, Ford's designers and technicians have never been short of creativity. This book tells their story, in their own words.

ISBN: 978-1-845849-86-3
Hardback • 25x25cm • 224 pages • 330 pictures

For more information and price details, visit our website at www.veloce.co.uk • email: info@veloce.co.uk • Tel: +44(0)1305 260068

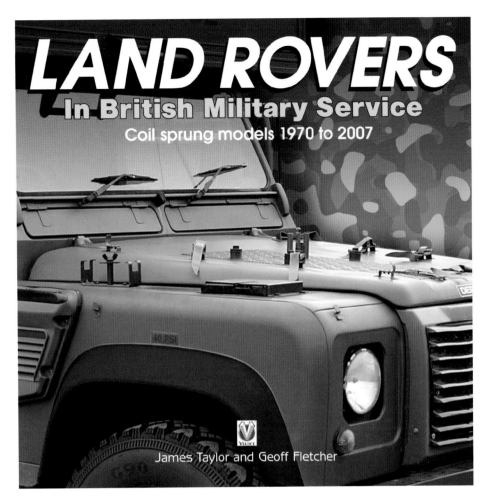

For anyone interested in the coil-sprung Land Rovers that have served (and still serve) with the British armed forces, this book is a must. It covers first-generation Range Rover and Discovery models, as well as the One Ten, Ninety and One Two Seven, their Defender successors and, of course the Wolf XD derivatives. Coverage deliberately ends at 2007 to respect current military sensibilities. This unique and extensively illustrated book describes and illustrates British military use and adaptations of these vehicles, and also contains comprehensive vehicle lists and contract details. The book is a sequel to *British Military Land Rovers, the leaf-sprung models*, by the same two authors (published by Herridge & Sons in 2015).

ISBN: 978-1-787112-40-7
Hardback • 25x25cm • 176 pages • 267 colour and b&w pictures

For more information and price details, visit our website at www.veloce.co.uk • email: info@veloce.co.uk • Tel: +44(0)1305 260068

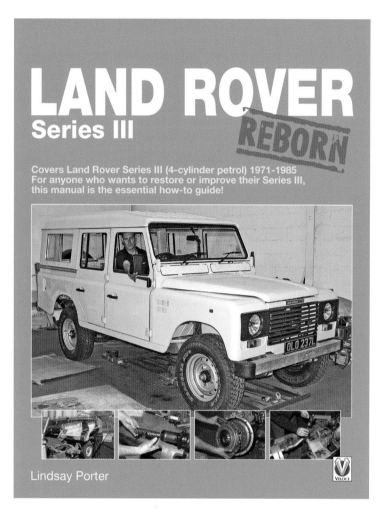

Developed from several years' articles in *Land Rover Monthly* magazine, this manual is the most detailed package of information available for anyone thinking of restoring, rebuilding or improving a Series III Land Rover.

ISBN: 978-1-845843-47-2
Paperback • 27x20.7cm • 256 pages • 1749 pictures

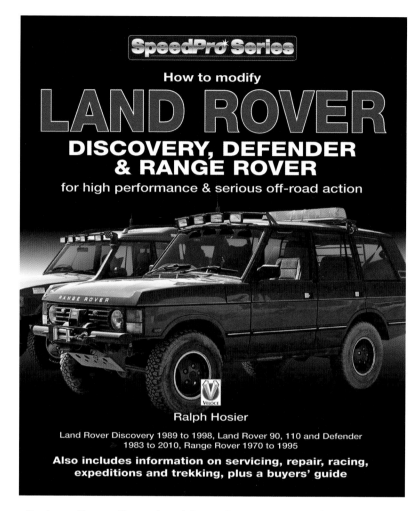

Buying a Range Rover, Land Rover Discovery or Defender can be just the start of a wonderful adventure. This book describes the options available to the owner, from big wheels and suspension lifts, under-body protection and tuning ideas, right up to how to convert the car into a high speed racer or an international expedition vehicle. With clear, jargon-free instructions, advice on events like family weekend green-laning, international expeditions and full-on competition, accompanied by colour photographs throughout, this is the definitive guide to getting the most from these exciting vehicles.

ISBN: 978-1-845843-15-1
Paperback • 25x20.7cm • 128 pages • 312 colour pictures

For more information and price details, visit our website at www.veloce.co.uk • email: info@veloce.co.uk • Tel: +44(0)1305 260068

Index

Aluminium in design 15, 52, 74, 97, 123, 134, 187, 188, 191
Alvis 38, 41, 45, 46
Anglesey, Red Wharf Bay 8, 11, 209
Arbuckle, David 71, 96
Aston Martin 41, 56, 137, 169
Austin-Rover 71, 72, 83, 88
Axe, Roy 71, 72, 95-97, 116, 170

Bache, David 6, 21-25, 32, 40, 41, 45, 46, 51, 55, 56, 71, 72, 78, 86, 96, 134, 224, 226, 230
Bangle, Chris 131, 133
Bannock, Graham 41
Bartlam, Dick 73, 84, 86, 89
Barton, Ken 24, 40
Barton, Tom 14, 16, 21, 24, 33, 38, 50, 70, 72, 73
Bashford, Gordon 14, 26, 28, 41, 46, 50
Beckles, Earl 161, 178
Beech, Ian 40, 41
Bertone 102, 103, 204
BMC company 25, 34, 46, 224, 226
BMW 56, 97, 105, 107, 110, 112-116, 119-122, 131, 133, 136, 137, 139, 142, 147, 153, 160, 170, 184, 187, 197, 226, 227
Boyes, Sandy 171, 173, 177
Boyle, Robert 10, 11, 14, 16, 24, 28, 29
Brisbourne, David 7, 76
Brisbourne, Kim 7, 76, 88, 134
British Aerospace (BAE) 83, 95, 97, 99, 112, 114, 122
British Leyland 10, 16, 32, 50, 52, 55, 56, 59, 62, 66, 75, 83, 96, 159
Brogan, Mike 76, 86, 94
Buffery, Martin 161
Butler, Mark 7, 154, 156, 171, 174, 177, 191, 200, 205, 206, 210, 214, 215

CAD in design 103, 154, 161, 178, 226-228, 230
Callum, Ian 142, 159, 226, 227
Callum, Moray 159
Canley studio 40, 56, 71, 72, 75, 95-98, 109, 111, 122, 126, 127, 131, 170, 178, 226
'Capricorn' project 72, 73
Car of The Year Award 30

Carbodies, Coventry 79, 80
'Challenger' project 98-101, 210, 223
Clay modelling 24, 32, 40, 46, 48, 73, 74, 76, 86, 87, 97, 107, 123, 126, 134, 145, 146, 161, 179, 198, 220, 226
Coldham, Charles 40, 57, 151
Colour and trim developments 16, 17, 19, 25, 40, 59, 64, 65, 73, 75, 76, 88, 91, 92, 94, 105, 112, 119, 123, 134, 135, 175, 204-207, 210, 216, 220-222
Conran Design 88, 89, 91, 92, 94
Coventry University 6, 96, 121, 170
Cowley plant, Oxford 122
Crathorne, Roger 7, 15, 16, 50, 64, 65, 68, 112, 131
Crompton, Geoff 40, 41, 43, 45, 54
Crompton, Pauline 40, 88
Crowley, Peter 7, 76, 98, 100, 164, 166

Day, Graham 83
DC100 concept 204, 210, 211, 223, 224
Defender
  1990-2002 11, 65, 67, 71, 98-101, 117
  2002 updates 164
  2007 L316 164-168
  2012 updates 206
  New Defender 223, 224, 227
  Special Editions 100, 101, 118, 168, 207-209, 224, 225
Design Council Awards 93
Discovery
  Series 1 59, 73, 83-94, 97, 101, 120, 125, 129
  Series 2 125-130, 142, 143
  Series 3 L319 136, 142-153, 162, 197
  Series 4 'LR4' 151-153, 182, 183, 187, 194, 196, 200, 204, 206
  Series 5 L462 199-207, 216, 220
  'Vision concept' 197-199, 204, 205
Discovery Sport 196-199, 204
Drayton Road studio, Solihull 71, 72, 74, 76, 77
Dunsfold Collection, Surrey 13, 32, 75, 116, 131

Eastnor Castle test circuit 33, 43, 50, 69, 112, 130
Edwards, John 184, 186, 217, 220, 221

Evans, David 7, 72, 74, 76-78, 82, 88, 92, 101, 102, 225
Evoque 170, 178-181, 185, 190, 194, 197, 199, 204, 205, 212, 228

'Farmers Friend' project 131
Fiat Campagnola 25, 59, 62
Ford 11, 41-44, 46, 52, 66, 83, 96, 121, 136, 137, 139, 142, 153, 159, 160, 164, 169, 170, 176, 181, 186, 188, 196, 197, 204, 226, 227
Frascella, Amy 7, 197, 205, 216, 226
Frascella, Massimo 7, 197, 200, 204, 216, 228
Freelander
  Series 1 CB40 107-112, 116, 120, 121, 127, 139, 140, 141, 159-161, 197
  Series 2 L359 158-163, 178, 197

Gaydon design studio 40, 122-124, 128, 131-133, 137, 142, 178, 198, 204, 226-228
Gaydon Research and Engineering Centre 10, 66, 120, 122-124, 137, 228, 229
General Motors (GM) 10, 50, 83, 197
Gilroy, Tony 67-69, 72, 73, 80, 83, 84, 88, 92, 102
Goddard, Arthur 14, 16

Halewood plant, Merseyside 158, 162, 182
Hamblin, Richard 71, 109, 121
'Heartland' project 125, 127, 143, 223
Henstridge, Sean 161
Hill, Maureen 7, 40, 45, 72
Hodgkinson, Mike 66, 67, 79
Honda 72, 97, 108-110, 112, 114, 120, 159, 170, 227

Ingeni design studio, Soho 169, 170
Issigonis, Alec 30, 112, 224

Jaguar 40, 41, 55, 57, 71, 83, 86, 101, 118, 136, 137, 139, 142, 169, 170, 187, 196, 197, 212, 217, 226-228
'Jay' project 73, 76, 83-94, 102, 103, 109, 110
Jeep cars 41, 42, 43, 50, 83, 127, 159

Jones, Mick 7, 72-75, 80, 88, 97, *Judge Dredd* film project 116

Keatley, Joanna 171, 175-177, 210
King, 'Spen' 10, 33, 41, 43, 45, 46, 50, 52, 66, 72, 109, 226

Land Rover
  127in, 130in 71, 98
  Forward Control 101in, 109in 32-36, 38, 50, 66
  'HiCap' 67, 86
  HUE 166 'Huey' 14, 185, 208, 209
  Ninety 67, 70, 71, 74, 100, 101
  One Ten 16, 38, 66, 68, 69, 71, 101, 226
  Series I 80in 11-18, 134, 185
  Series I 86in 16-25
  Series II, IIA 24-26, 30, 33, 35, 41, 49, 50, 56-58, 65
  Series III 16, 56-59, 65-67, 69
  Stage 1 V8 66-68, 71
Le Grice, Oliver 7, 96, 97, 122, 123, 178, 210, 229
Lewis, Graham 40, 41
Llama forward control truck 73-75, 99, 109
Lode Lane 38A design studio 75-77, 93, 102, 109, 220
Lode Lane, Bache design studio 21, 24, 32, 40, 41, 46-48, 56, 60, 71, 76, 77, 86, 88, 89
Lode Lane plant, Solihull 9, 10, 12, 14, 20, 21, 24, 27, 30, 55, 67, 76, 117, 122, 139, 158, 164, 182, 185, 199, 209, 216, 220, 221
Loker, Harry 24, 26-28
Longbridge plant 32, 56, 71, 122, 131, 137
LRX concept 170-177, 180, 182, 185, 205, 210

Mackie, George 10, 38
Mann, Harris 56, 71, 72, 95
Mays, J 137, 159
McGovern, Gerry 57, 71, 96, 110, 112, 115, 131, 137, 169, 170, 174, 178, 180-182, 184, 199-204, 206, 210, 212, 217, 223, 226, 227, 230

Miller, Geof 33, 43, 50, 64
Mobberley, Alan 7, 76, 89, 91, 92, 98, 108, 125, 128, 129, 134, 167
Monteverdi Range Rover 78, 79
Morris, Bill 70, 73, 78
Morris, Norman 7, 40
Muncaster, Matt 73, 76
Musgrove, Harold 32, 72, 83, 96

Ozozturk, 'Memo' 72, 76, 86, 101, 112

PAG (Premier Automotive Group) 137, 142, 159, 169, 184
Paint colours 15, 16, 19, 25, 54, 65, 73, 80, 101, 117, 164, 168, 193, 206, 208, 214, 220, 224, 227
'Pathfinder' project 104, 107-109, 197
'Pegasus' project 73, 103
Pininfarina (also Farina) 23, 29, 56, 102
Plastics in design 40, 49, 53, 54, 56-58, 92, 98, 103, 104, 166, 188
Pogmore, Col. Jack 25, 29, 33
Poole, Tony 23, 24, 40, 41, 48-50, 56, 57, 67, 71, 72, 76, 78, 86, 230
Pressed Steel Company 7, 55, 56
Product Planning 41, 72, 80, 83, 84, 86, 93, 98, 107, 110, 120, 127, 139, 147, 153, 154, 160, 161, 197, 215
Purkis, Geoff 7, 40

Range Rover 16, 38, 40, 71, 182, 186
  Series 1 41-55, 59, 63-65, 72, 78-82, 104, 125, 128, 137, 221
  Series 2 P38A 65, 101-106, 112, 116, 117, 125, 127, 131, 182
  Series 3 L322 131-139, 143, 162, 186, 187, 191, 226
  Series 4 L405 186-196, 200, 216, 220
  SV Coupé 221-223
Range Rover Sport
  Series 1 L320 153-155, 161, 182, 186, 194, 195
  Series 2 L494 186-197, 199, 200, 212, 213, 218-221
Range Stormer concept 155-159
Reichmann, Marek 131, 137

Reitzle, Dr Wolfgang 112, 116, 131, 134, 136, 137, 139, 142, 184
Road-Rover 25-29, 41, 45, 46, 228
Rootes Group 10, 40, 220
Rover Cars
  200 97, 107, 109, 112, 116, 121
  400 97, 109, 116, 121
  600 97, 121
  800 97, 103, 104, 125, 127
  M-Type 10
  P2 14
  P3 11, 14, 15, 24
  P4 21, 23, 24, 26, 29, 32
  P5 21, 23-25, 28, 29, 32, 41, 46, 56
  P6 24, 25, 30-32, 41, 49, 50, 52, 54, 56, 67
  P8 46, 49, 50, 55, 56
  P9 46, 55, 56
  SD1 (P10) 32, 56, 67
  SD2 56
  T3 Turbine car 23, 56
Rover Company 8-10, 16, 20, 24, 40, 46, 50
Rover Group developments 83, 95, 107, 114, 122, 159
Rover 'Oden' project 97, 104, 107, 108, 197
Rover Special Projects 109
Royal College of Art, London 6, 72, 96, 121, 170
Rubery Owen 'Rostyle' wheels 52
Ryder report 62

Saddington, Dave 7, 71, 96, 97, 143, 145, 178, 197, 199, 226, 227
Sampson, Mike 7, 76, 84, 85, 87, 125, 126, 129, 132, 136
SD5 project 56-62, 223
Sheppard, Alan 7, 56, 74, 76, 86, 87, 103, 112, 134, 226, 230
Simmons, Phil 7, 96, 97, 131-133, 137, 181, 182, 184, 185, 189, 200, 226, 228
Sked, Gordon 71, 95-97, 116
Slater, Joanne 191
Smith, Len 40, 56, 97, 123
Solihull – see Lode Lane plant
Special Projects 34, 38, 39, 72, 78, 80
Speth, Dr Ralf 184, 186

Spindler, Kevin 40, 41
Stark, John 7, 76, 91, 96
SVO 38, 72, 75, 100, 118, 119, 168, 217-223

Tata Motors 169, 178, 184, 186, 227
Thomson, George 73, 76, 84, 97, 101, 102
Thomson, Julian 171, 172, 177
Tickford 16, 17, 26
Towns, William 40, 41
Triumph cars 24, 25, 30, 50, 52, 54-56, 65, 67, 68, 71, 72, 97, 226

Underwood, Frank 24, 25, 40
Upex, Geoff 7, 96, 97, 110, 114, 116, 120, 121, 136, 137, 139, 142, 145, 151, 153-156, 159, 161, 169, 170

Vauxhall cars 66, 83, 84, 96, 226
Velar 2017 185, 212-217
Velar prototypes 1969 47, 51, 53, 54, 213
Volvo cars 32, 49, 55, 137, 147, 160, 161, 170, 197

Wade, Chris 7, 40, 41, 55, 57
Waterman, Jez 171, 174, 177
Watkins, James 161, 178
Wheel, Andy 7, 96, 143-146, 149, 153, 166, 167, 178, 181, 227
Whitley plant, Coventry 71-73, 76, 96, 97, 101, 102, 142, 170, 229
Wilks, Maurice 8, 10, 11, 14-16, 21, 23, 24, 26-28, 30, 32, 38, 50, 226, 228
Wilks, Nick 10, 43
Wilks, Peter 10, 33, 38, 45
Wilks, Spencer 8, 10, 16, 25, 38, 220, 226, 230
Willys Jeep (see also Jeep cars) 10, 11, 25
Woodhouse, David 96, 107, 112, 116, 137
Woolley, Richard 7, 96, 109, 137, 142, 159, 161, 170, 186, 187, 190, 193, 197, 200, 210, 217, 219-221, 223, 224, 227, 229
Wyatt, Don 7, 76, 101, 102, 109, 131, 133

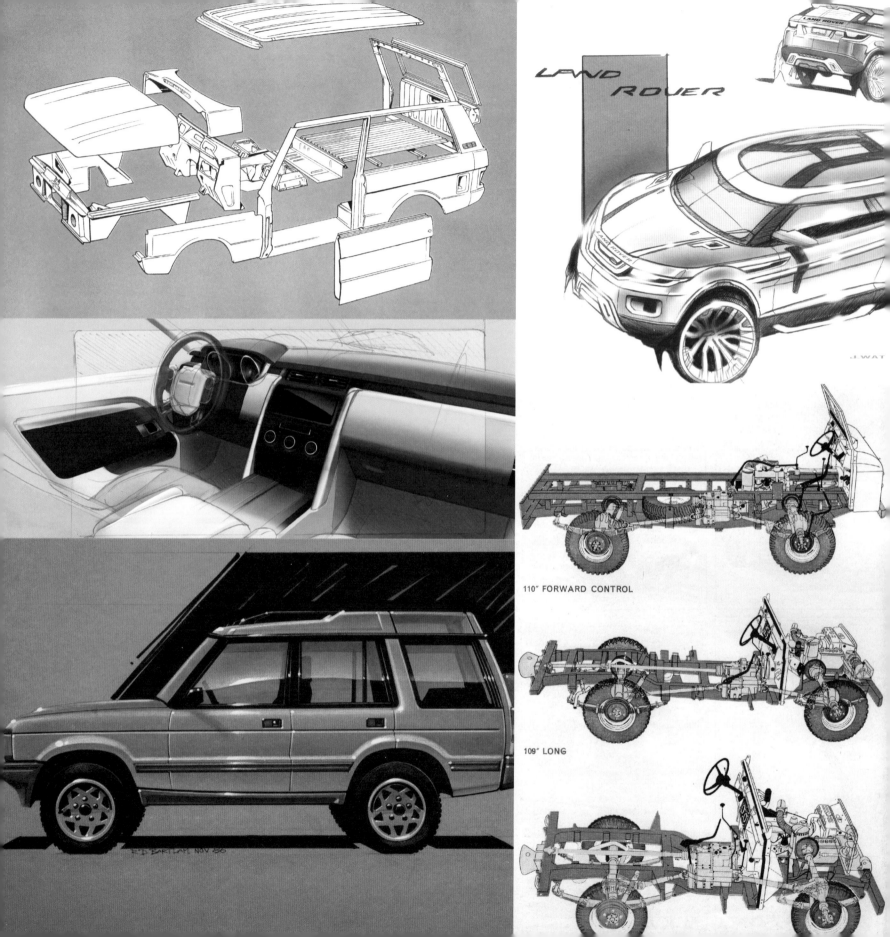

110" FORWARD CONTROL

109" LONG